Aline Piccini Roll
Fernando Rutz
Victor Fernando Roll

Quail: production, management and nutrition

Aline Piccini Roll
Fernando Rutz
Victor Fernando Roll

Quail: production, management and nutrition

ScienciaScripts

Imprint

Any brand names and product names mentioned in this book are subject to trademark, brand or patent protection and are trademarks or registered trademarks of their respective holders. The use of brand names, product names, common names, trade names, product descriptions etc. even without a particular marking in this work is in no way to be construed to mean that such names may be regarded as unrestricted in respect of trademark and brand protection legislation and could thus be used by anyone.

Cover image: www.ingimage.com

This book is a translation from the original published under ISBN 978-613-9-62530-7.

Publisher:
Sciencia Scripts
is a trademark of
Dodo Books Indian Ocean Ltd. and OmniScriptum S.R.L publishing group

120 High Road, East Finchley, London, N2 9ED, United Kingdom
Str. Armeneasca 28/1, office 1, Chisinau MD-2012, Republic of Moldova, Europe
Printed at: see last page
ISBN: 978-620-7-73015-5

index

Preface 2

1. General considerations 3

2. Types of breedings 16

3. Experimental methods for the evaluation of quail production. 20

4. Evaluation of egg quality 24

5. Quail management 35

6. Nutrition and food 48

7. Anatomy of the quail 69

8. Production factors related to quail welfare. 73

9. Bibliography 79

10. Authors 89

Preface

Quail farming is the raising of quail for the production of eggs or meat. On these farms, producers and researchers conduct production or experiments following methods and procedures that guarantee the quality of the products, the reliability of the results and the replicability of the studies.

This activity is relatively recent, mainly in Brazil, so many people are still unaware of it. Although Cotorniculture in Brazil does not have the same consolidated tradition of European countries, we already have great producers and high level researchers in this area of scientific knowledge.

We wrote this book as a guide, a support tool, intended for students in the zootechnical, veterinary and agronomic fields, as well as for technicians and producers, but we hope it will be read by all those who work with and like these birds.

This book will help to describe the main points of quail breeding for fattening and/or laying in relation to productive, reproductive and nutritional management.

With this book the reader will be able to better understand from the characteristics of this bird to how to use the techniques of scientific methodology in quail research.

1. General considerations

1.1 Cotorniculture

Quail farming, whether for meat or egg production, is a sector of poultry farming that is in full development, with productivity and profitability that have attracted the attention of producers. Among the factors that contribute to its success are the rapid development of quail, the short interval between generations, precocity in production and sexual maturity, as well as longevity in high production (14 to 18 months) and rapid return on investment (Pinto et al., 2002; Murakami and Furlan, 2002).

However, quail meat is little known in Brazil because it is a recent commercial exploitation, and because it is considered a noble meat that is difficult for the population to access (Reis, 2011).

Quails belong to the same family as chickens and partridges, i.e., the family of the Fasianidae (Phasianidae) and the subfamily of the Perdicinidae and are native to Europe, Asia and North Africa (Pinto et al., 2002).

Three different types of quail have been exploited in industrial breeding since the 1990s. They are: *Coturnix coturnix japonica* (japonica quail); *Bobwhite Quail* (American quail); and *Coturnix coturnix coturnix coturnix*, (European quail). These birds differ in size, precocity, egg coloration (white or painted), laying rate and feather coloration.

1.2 Comparison between quail lines

Studies indicate that European quail selected for fattening have higher egg mass when compared to quail for egg production (Singh and Panda, 1987). However, there is little information regarding feed conversion per dozen eggs, egg quality and feed consumption, making it difficult to understand their productive potential in terms of their dual suitability (Mori et al., 2005).

The japonica and European quail are phenotypically similar, but the European quail have a higher live weight (between 200 and 300 grams), more vivid brown coloration and a temperament characteristic of slaughter animals. They are also notably calmer, whether raised on the floor or in cages, in spite of having similar age of sexual maturity (Rezende et al., 2004).

Oliveira (2003) reports that European quail consume about 36 grams/day of feed in the production phase and that feed costs represent about 70% of the total. However, studies to confirm these data are still scarce, as far as European quail are concerned, since most of the research and results found in the literature refer to japonica quail.

The carcass yield of quail represents about 72% of their live weight. This result is one of the highest among birds, in addition to having relatively large eggs in relation to their size (Murakami and Furlan, 2002). The same authors show that the regulation of feed intake occurs by the birds' demand and energy density of the diet when the birds are fed ad libitum.

The precocity of cutting quails is achieved due to the high feed consumption in their initial phase of life, presenting higher weight and growth rate than laying quails (Marks, 1991).

For a better development of breeding, research is providing a better knowledge of the available lines to establish their zootechnical levels (Móri et al., 2005).

Currently, in Europe and the United States, the most prevalent lines are dual-purpose, i.e., those used for egg and meat production (Banerjee, 2010). Although in many countries quails are exploited for dual purpose (Jones et al., 1979, Baumgartner, 1994), in Brazil their exploitation is still almost exclusively for egg production, with only males that were wrongly classified in the sexing process performed at one day of life and females at the end of their productive life being slaughtered. In the latter case, these are animals without a fixed age pattern, generally older than 52 weeks, which for this reason have impaired carcass characteristics (Reis, 2011).

Although quail farming in Brazil is generally carried out in small farms, it is possible to obtain a closed cycle of breeding, incubation and production in the same place. However, in this form of breeding, the risks of bacterial contamination are higher. One option for the prevention of these risks can be the specialization of the sectors with the implementation of aviaries divided into sectors, such as, for example, selection sectors, multiplication sectors and commercial production sectors.

1.3 Origin

The quail is a bird of beige plumage with small black and white stripes. It is a very old migratory bird in Europe and was first taken to Asia (China and Korea) and later to Japan.

The European quail was introduced to Japan in the 11th century via China. Early accounts say that the Japanese began to breed them because of their song. In Japan, studies and crosses were carried out that made possible the appearance of the japonica quail destined for meat and egg production (Reyes, 1980). The Japanese, around 1300 A.D.   ,   began   to breed quail for meat and eggs (Reyes, 1980).

domestication of the species based on the melodious song of the males, and after several attempts, around the 20th century, the commercial breeding of the species was promoted with mass production in cages.

From the Coturnix coturnix subspecies, a wild subspecies of European origin, the Japanese and Chinese, after several crosses, managed to produce the current quail, the *Coturnix coturnix japonica,* also known as the Japanese quail.

Due to its high precocity, productivity and fertility, as well as the low space requirements for breeding and ease of transport due to its small size, quail became one of the main food sources for the Vietnamese during the war against the United States.

In the 1950s, Italian and Japanese immigrants brought quail to Brazil and since then production has been consolidated, becoming an important alternative source of food, mainly eggs. The activity is a profitable alternative for the agricultural sector when carried out professionally. However, the great majority of the productions are of small properties, with some being able to be of closed cycle, with reproductive, incubation and finishing sectors.

Although there is still no specific line for laying and fattening in Brazil, in order to achieve better performance, producers end up bringing new lines from other countries. Quail meat is still considered exotic and has high prices. However, consumers have been changing their eating habits over the last few years and have become more accepting of this type of protein, which can promote increased production in the long term.

Normally, quail meat (Figure 1) found at the commercial level is produced by a few specialized companies. In this case, imported lines from Europe are used, with the capacity to reach up to 600 g of weight at a slaughter age of 60 days. In alternative breeding, there are male animals that were not used in reproduction and discarded females that have already finished their productive cycle.

Figure 1. Carcass of European dual-purpose quails (approximate weight 400g)

According to Pinto et al. (2002), the quails introduced in Brazil in the 1950s were very similar to those already existing in the country. However, they did not belong to the same family of the Tâmamidae, as is the case of the northeastern quail (*Nothura boraquira*), the minera (*Nothura minor*) and the common quail (*Nothura maculosa*).

Incubation of quail eggs lasts about 17 days with chicks on the first day of life weighing about 7.0 grams in Japanese quails and 10 to 12 grams in European quails, which corresponds to approximately 70% of the egg weight. At the end of four weeks, quails increase 10 times their initial weight (Albino and Barreto, 2003).

Adult females are heavier than males due to the development of the reproductive apparatus and liver. Because they start laying around 42 days of age, they are considered the birds with the earliest physiological development, producing up to 300 eggs in the first year of life weighing between 9.0 and 12.5 grams, which can represent up to 8% of the bird's weight (Belo et al., 2000). Eggs usually have pigments all over the shell surface with patches of various colorations in shades of brown (Albino and Barreto, 2003).

1.4 Nutritional value of meat and eggs

The nutritional properties of quail products are related to the quality and quantity of the nutrients contained in them. Quail need to receive in their diets all the essential nutrients in adequate proportions and quantities so that they can optimally express their productive and reproductive functions.

Nutrients are also responsible for the nutritional and sensory characteristics of meat and eggs. These nutrients can be divided into two categories, which are the nutritive (basic) constituents and the secondary constituents.

As nutritional components can be cited water, carbohydrates, proteins, minerals, vitamins and lipids. And as secondary constituents can be cited organic acids, enzymes, pigments, volatile compounds, essential oils, among others.

With respect to the nutritional and sensory characteristics of foods, each of the above-mentioned constituents is responsible for a specific characteristic (Teixeira, 1989):

Nutritional value → protein, carbohydrates, lipids

Color → enzymes, pigments

Flavor → organic acids, carbohydrates

Odor → essential oils, volatile compounds

Texture → pectins, proteins

1.5 Location of implementation and installations

Quail can basically be used for four types of breeding: day-old chicks, breeders, meat production (fattening) or egg production (laying).

Regardless of the type of breeding, in the choice of the location of the facilities it is important to consider some determining points such as, for example, the ease of supplying the property with inputs for poultry feed, the output of the production at the time of flock or egg removal, the proximity to the slaughterhouse, as well as the ease of access to the consumer market, among others.

Regions with very abrupt climatic variations should be avoided, as well as those with high incidences of flooding and very high relative air humidity rates. The site to be chosen, therefore, whenever possible, should be easily accessible, be supplied with good quality electricity and water, and should not be subject to windstorms or floods.

Normally, it is advisable for these farms to be located in rural areas. However, what determines this issue is the urban master plan (PDU) of the municipalities of each locality. Generally, it is the PDU that determines the type of activity that can be installed or implemented on the property or real estate chosen for the installation of the company. This, therefore, must be the first step to evaluate the implantation of a quail farm and the city council of the corresponding city must be consulted.

It should be noted that these requirements are indicated for those producers who want to start with commercial breeding. In this sense, it is important that the basic structure has a good availability of water and area, as well as a favorable climate.

In order to allow the birds to express their genetic potential to the maximum, it is recommended that the structure be built according to the environmental requirements of the quails, providing an optimal condition of comfort for them. Depending on the type of commercial or even domestic breeding, some equipment and installations will be necessary, which will be described below.

1.5.1 Open sheds on the sides

This type of house allows greater economy when installed in areas with mild temperatures, however, it requires temperature control during the winter or colder periods.

These installations require mesh on the sides, in order to prevent not only the birds from escaping from inside the house, but also to prevent the entry of other birds or predators from outside the breeding area.

1.5.2 Closed sheds on the sides

Sheds closed on the sides are more expensive than open sheds. In addition, they should not be too long or too wide to facilitate air circulation, and it is recommended to install several openings in the form of windows on the sides.

1.5.3 Roofs

Roofs are extremely important as they greatly influence the internal temperature of the aviary. There is a diversity of coverings and materials that can be used as roofing for the facilities. Clay tiles, for example, offer greater internal thermal comfort, however, they require greater expenditure with wood to support the roof. On the other hand, fiber cement tiles are lower cost, however, they increase the internal temperature.

In the aviary cover, other measures can be used to soften the internal temperature of the house on very hot days, such as painting the external surface of the roof with light colors, spraying water on the roof on very hot days, increasing the inclination of the roof, since it facilitates the dissipation of heat outside the installation when at steep angles.

However, in spite of being more expensive, there are nowadays tiles with efficient thermal insulation systems, with which greater production and welfare of the birds can be obtained, being possible in the long term to become the most profitable option for the producer.

1.5.4 Floor

The floor of the aviary can have a base of rustic cement or other type of material. Regardless of the type of floor used, it must have a small slope so that water does not accumulate inside the aviary.

1.5.5 Recria Cages

Rearing and rearing cages are used in the intermediate growth phase, where quail are housed at 15 days of age and are removed at 35 days of age when they are transferred to laying cages. It is also possible to house the birds directly in the laying cages when they reach 15 days of age.

1.5.6 Laying Cages

Laying cages allow a better nutritional, sanitary and productive control of the birds. For this purpose, standardized galvanized cages with dimensions of 100 cm x 30 cm are recommended, which house up to 30 birds. Cages with dimensions of 100 cm x 40 cm with up to 40 birds can have two or three partitions in cage battery systems with up to five or six tiers.

1.6 Equipment

This sub-index illustrates the necessary equipment in the facilities, such as, for example, drinking troughs, feeders, hoods, fans, foggers, bedding, cages, protection circles, among others.

1.6.1 Drinking fountains

Drinkers can be of the infant type (the same used in broilers), the cup type or the nipple type. The choice and use will depend on the phase and the rearing system.

In the rearing system on litter in the rearing and rearing phase, the drinkers can be of the infant model, which have a size variation from 800 ml to 7.5 liters, and can be used for broiler rearing. However, the use of a hose inside the compartment is recommended to avoid possible drowning of the chicks, as shown in Figure 2.

Figure 3 shows a model of a pendulum drinker. It can be used in the production of fattening quails raised on litter. In cage rearing systems the drinkers are usually of the nipple type.

Figure 2. Adapted broiler drinker with a capacity for 5 L of water.

Figure 3. Pendulum type drinker for larger birds

The most commonly used drinkers are the nipple model, due to the ease of sanitation, resulting in better water quality, less waste and greater control in the administration of medicines. However, it is necessary to be careful with the maintenance of the nipples, because they can get stuck promoting undesired water losses inside the aviary. In the same way, the tubing should also be checked periodically to avoid accumulations or air inlets that can cause lack of water in the nipples.

1.6.2 Feeders

In the industry there are several models and shapes of feeders, which can be manual or automatic, for children or adults, with a series of sizes and colors. Regardless of the model, feeders should be arranged in the aviary in sufficient numbers to provide for all birds and should always be well stocked and adjusted to the age and height of the birds. Figure 4 shows a model of a tubular feeder with a capacity of 5 kg of feed.

In the case of cages, the feeders are generally made of galvanized metal and are arranged at the front of the cage.

Figure 4. Tubular trough with capacity for 5 kg of feed.

1.6.3 Bells

Hoods are heaters that serve as a source of heat for the chicks in the first weeks of life or whenever necessary.

Hoods are the most commonly used heat sources as they can be used with either natural gas or liquefied petroleum gas (LPG), as well as electric or wood. In the industry there is a diversity of models with different shapes and automatic temperature controls being most of them very functional and resistant with low maintenance rates.

Gas hoods operate by means of a conventional gas burner and heat is supplied to the chicks through convection and conduction. Although there are a variety of hood models on the market of different sizes (for 125 or 2500 birds), they are usually rounded in shape and can each provide heat for up to 500 chicks (Figure 5). The installation is considered easy and simple, usually installed on the roof so that they are close to the birds, promoting heating within the radius of action. Control of the heat supply is provided by manual or automatic adjustment of the desired temperature.

Figure 5. Gas hoods as a heating system

Gas hoods with refractory ceramic plates are considered very good because they allow heat supply by radiation, where the ceramic plate becomes incandescent when it reaches the desired temperature and thus transfers the heat produced. In this case, these hoods can be installed at a higher height compared to conventional hoods, as the temperature is distributed more evenly and can provide heat for up to 800 chicks within the radius of action of the hood. The possible breakage of the ceramic plate during the handling of the equipment is a disadvantage compared to the previously mentioned hoods.

Infrared gas hoods, in which the burning of the gas promotes heating with the use of radiation, have highly heat-resistant metal burners. The main difference between the forms of heat transmission is that in the radiation system the highest temperature produced is within the comfort zone of the birds, while in the convection system the heated air tends to rise to the higher zones, which promotes layers of air with uneven temperatures.

2. Types of breedings

In general, there are basically four types of broodstock: day-old chicks (Figure 6), breeders, broiler farms and egg farms.

The increase in consumption has raised quail farming to higher levels. In 2014, 20.34 million quails were registered in Brazil, which represented an increase of 11.9% compared to that recorded in 2013. From the analysis of the records from the years 2005 to 2014, the historical growth of the number of quails in that period is observed, which represents the constant increase in the production of that species. The highest concentration of production (78.2%) was observed in the Southeast Region, with the largest producer of quail being the state of São Paulo, with 54.5% of the total produced in Brazil (IBGE, 2014).

Quail egg production in 2014 was 392.73 million dozens, which represented an increase of 14.7% compared to 2013 records. Quail egg productions are mainly located in the Southeast Region (82.1%). The State of São Paulo stood out as the largest national producer (59.3%), followed by the States of Espirito Santo (11.5%) and Minas Gerais (9.6%). The municipalities of Bastos (SP), Lacri (SP) and Santa Maria de Jetibà (ES), which have the largest number of animals, also had the highest quail egg production in 2014 (IBGE, 2014).

The advantages of quail farming include rapid growth with early sexual maturity, high productivity, high prolificacy, large number of birds that can be housed in a small space, longevity of production, low initial investment and quick financial return, in addition to the exotic flavor of their meat.

Figure 6. Day-old double-purpose quail chicks

2.1 Fattening farm

The line most commonly used for meat production is *Coturnix coturnix coturnix coturnix*, also known as European quail (Figure 7). With the increase in world demand for meat, it is necessary to investigate alternatives that can meet the new demands for animal products, and quail production is an interesting option (Móri et al., 2005).

Figure 7. European dual-purpose quail *Coturnix coturnix coturnix coturnix* (laying

and meat)

In this sense, quail meat production is shown to be an efficient alternative, as less investment and space are potentially required for the facilities (Figure 8). On the other hand, the production of waste is relatively lower compared to other breeding operations. However, knowledge about performance, nutritional requirements and suitable genetic materials is still scarce, directing quail production mainly to egg production.

Figure 8. Exploration of quails on bedding intended for fattening.

2.2 Egg production farms

The main line used for egg exploration is *Coturnix coturnix japonica*, also known as japonica quail (Figure 9).

Laying poultry enables a quick return on invested capital due to the early maturity of the birds. On the other hand, it is of great interest to the consumer as a source of food rich in vitamins, minerals and high quality proteins (Seibel et al., 2010).

Quail eggs contain about 6.5 mg of protein, 40 mg of vitamins, 112 mg of phosphorus, 1.85 mg of iron and 31 mg of calcium in 50 g of product and are therefore an excellent source of nutrients (Redder, 2005).

The incubation process of quail eggs normally lasts 17 days, with chicks weighing an average of 7.0 grams on the first day of life in Japanese quails and 10 to 12 grams in European quails, which corresponds to approximately 70% of the egg weight.

At four weeks of age, quails increase 10 times their initial weight (Albino and Barreto, 2003). Females as adults are heavier than males due to the development of the reproductive apparatus and liver.

Because they start laying at approximately 42 days of age, quails are considered the birds with the earliest physiological development and can produce up to 300 eggs in the first year of life with weights ranging from 9.0 to 12.5 g, which can represent up to 8% of the bird's weight (Belo et al., 2000). Eggs usually have pigments all over the shell surface with patches of various colorations in shades of brown (Albino and Barreto, 2003).

Figure 9. Quail farming in a battery of galvanized cages for egg production.

3. Experimental methods for the evaluation of quail production

of quail

Evaluation of quail production can be used to monitor flock performance. Variables that can be measured include bird weight, feed consumption, egg production and weight, and feed conversion, among others. These variables are detailed below.

3.1 Body weight

The weight of the birds (Figure 10) should be controlled to monitor the development and uniformity of the flock, as well as to determine the possible presence of diseases from the observation of reduction or exaggerated weight gain in some quails or in the flock.

Figure 10. Individual bird weighing

3.2 Feed consumption

Feed consumption data are extremely important for the producer, since this variable allows to observe if the birds are within the stipulated standard for the line, sex and age of the quails. It can also be a tool to detect possible nutritional or feed formulation problems.

Feed consumption can be calculated using the following formula:

CPP = PFP - S

being,

CPP: total feed consumption in the period (g);

PFP: feed fed in the period (g);

S: feed leftovers (g) collected from each cage, at the end of the period (usually one week or one month).

The average daily feed consumption per bird can be calculated on the basis of total consumption using the following formula:

CMP = (CPP/y)/x

being,

CMP: average feed intake (g/bird/day/cage);

y: number of birds housed per cage;

x: number of days in the period.

3.3 Egg production

Egg production should be recorded daily to obtain the total number of eggs produced by the flock or by the experimental unit in case of animal experimentation during the period evaluated.

The calculation to obtain the percentage of eggs produced in each period can be done using the following formula:

Production (%) = (THx100)/y

Being:

TH = total eggs produced during the period;

y = number of eggs expected at 100% production during the period evaluated.

3.4 Weight of eggs

At the end of each production or experimental cycle, the eggs produced should be identified, collected and weighed (Figure 11). This weighing serves to monitor the productive pattern of the flock.

To verify the average weight of the eggs we can use the following formula:

PMH (g) = PTH/NTH

Where:

PMH: mean weight of eggs;

PTH: total weight of eggs produced during the period;

NTH: total number of eggs produced in the period.

Figure 11. Manual weighing of eggs

3.5 Feed conversion per kilo of eggs

Feed conversion (FC) is obtained from the ratio between total feed consumption and kilograms of eggs produced in the period according to the following formula:

CA = feed consumption/egg weight

Verification of kilograms of eggs produced can be done from the average weight of eggs multiplied by the production rate obtained in the desired or observed period. In the case of fattening quails the CA can be calculated using the following formula:

CA = average feed consumption/average weight of quails

For the interpretation of the results, we consider that an optimal CA is the one that is closest to 1 (one), that is, consumes one kilo of feed and produces one kilo of eggs or meat.

Using the above formula as an example, considering an average feed consumption of 1050.0 grams and an average egg weight of 300.0 grams during a given period, the CA would be 3.5, which means that, in this hypothetical example, the birds would need to consume 3.5 kg of feed to produce one kilo of eggs.

4. Evaluation of egg quality

Egg quality can be evaluated internally and externally. External quality comprises the evaluation of egg weight, specific gravity, size and diameter. Internal quality is evaluated by albumen weight and height, yolk coloration and weight, as well as sensory analysis. Chemical determinations can be carried out to evaluate moisture, protein, ethereal extract (lipids), ash and fiber contents, among others.

Egg quality can be influenced by many factors, such as nutrition, health, environment, genetics and quail management.

4.1 External quality

4.1.1 Weighing of eggs

Egg weighing (Figure 12) in commercial hatcheries is carried out to monitor flock performance and production. The success and participation in the consumer market depends on the quality of the eggs produced, which mainly includes characteristics such as yolk and shell color (Xavier et al., 2008).

However, the Brazilian Ministry of Livestock Agriculture and Supply (MAPA) through Normative Instruction (IN) n° 05 of 02/14/2017 - Art.64 - Single Paragraph determines that the production of quail eggs be dispensed from the weight grading stage. Therefore, eggs are marketed in cartons, usually with 30 eggs weighing about 11g each, unlike hens' eggs, which are graded and marketed by classes according to their weight.

The average weight of dual-purpose quail eggs is close to 11 grams, but in extreme cases can weigh up to 17 grams (Figure 12). It is important to note that the weight of quail eggs, as in other species of laying birds, varies according to dietary and environmental management, in addition to the age of the bird. Egg weight increases with advancing age (Roll et al., 2009) while eggshell quality worsens (Akyurek and Okur, 2009) and the Haugh unit value decreases (Fletcher et al., 1983).

Figure 12. Egg weighing in dual-purpose (European) quails.

4.1.2 Specific gravity

Specific gravity can be used as an indirect measure of eggshell quality (Baiâo and Cançado, 1997). However, in order to carry out this analysis it is important to take care in the collection and processing of these eggs so as not to break or crack them, as they will then be disqualified for this measure.

For the evaluation of specific gravity, the eggs are placed in buckets with saline solutions, from lower to higher concentration of sodium chloride (NaCl), with an interval of 0.004, which can vary from 1.046 to 1.090, totaling 12 solutions (Roll, 2012).

Prior to each evaluation, densities are determined with a densimeter. In this evaluation, the eggs are removed by floating in the water and pointing to the respective density value corresponding to the solution in the container (Figure 13). The specific gravity should be measured on the day of laying, since cracked eggs or eggs with soft shells should not be analyzed.

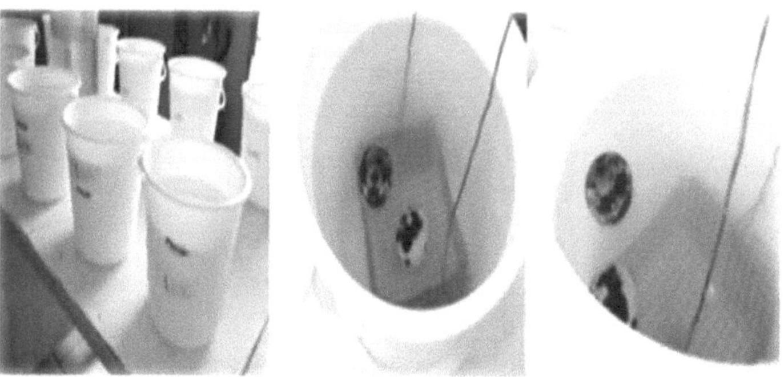

Figure 13. Cubes with different densities as a function of salt concentration for the evaluation of egg specific gravity.

4.1.3 Shell weight and thickness

The eggshell can be considered as the packaging of the egg contents. In this sense, it must be sufficiently resistant to the impacts that occur until it reaches the final consumer, such as laying, collection, grading and transport. Shell quality can be influenced by a series of factors that can be related to inadequate storage, age of the quail, rearing environment, nutritional management, among others.

The shell can represent from 9 to 12% of the total egg weight, being the first line of defense against microbial contamination. In addition, eggshells, when resistant, represent a more efficient protection of the internal contents of the eggs. On the other hand, regardless of the coloration or pigmentation of the shell, it is important that it is clean, whole and without deformations in order to promote this protection.

After breaking each egg for internal quality analysis, the shells should be washed to completely remove the albumen adhering to the inner membrane. They should then be placed in a forced air oven at 60°C for 24 hours to promote shell drying. After drying, the shells can be weighed with a digital balance preferably with an accuracy of 0.01g. A micrometer can be used to measure the shell thickness (Figures 14 and 15).

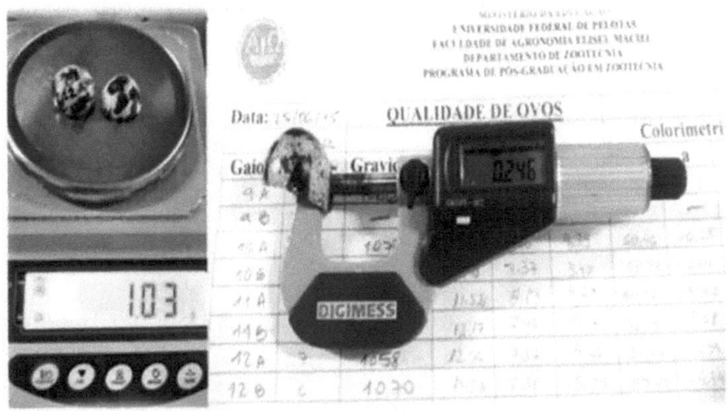

Figure 14. Weighing and measurement of shell thickness with digital micrometer.

Figure 15. Micrometer for shell thickness assessment

4.2 Internal quality

4.2.1 Albumen height

The contents of the egg are deposited on a container with a level surface (glass plates, dishes) and the albumen height is measured using a ruler specifically designed for this purpose.

4.2.2 Haugh Unit

The Haugh unit is a measure used to evaluate egg quality which, in addition to being considered the best evaluation parameter, is a measure of egg albumen quality.

Water and carbon dioxide losses through the shell pores are lower for quail eggs, when compared to chicken eggs, due to the greater thickness of the shell membranes. The higher the albumen height, the higher the Haugh unit and the better the egg quality.

The Haugh unit (HU) is obtained from the logarithmic relationship between egg weight and albumen height according to the formula exemplified below (Silva et al., 2000).

UH = 100log (H + 7.57 - 1.7W -3)^{07}

Being:

H = height of thickened albumen (mm)

W = weight of egg (g)

4.2.3 Yolk coloration

Two methods can be used to evaluate yolk coloration. The first is the use of a colorimetric fan in which the color of the yolk is visually compared with the existing colors in the fan, which contains scales of shades ranging from one (yellow) to fifteen (orange), as can be seen in Figure 16.

The second method uses a tri-stimulus colorimeter (color, hue and brightness) that must be previously calibrated on a white surface. With the use of the colorimeter the values of chroma L, which gives the brightness with a variation from white (L = 100) to black (L = 0), of chroma a*, which characterizes the coloration from red (+ a*) to green (-a*) and the values of chroma b*, which characterizes the coloration from yellow (+ b*) to blue (-b*) are evaluated.

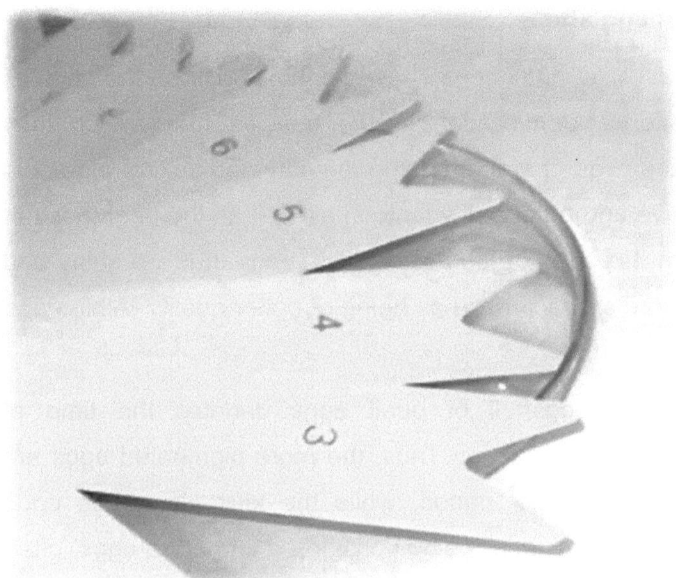

Figure 16. Colorimetric range for the evaluation of yolk color.

4.2.4 Yolk and albumen weighing

Weighing of the yolk and albumen can be done with the aid of an albumen separator and a digital scale (Figure 17).

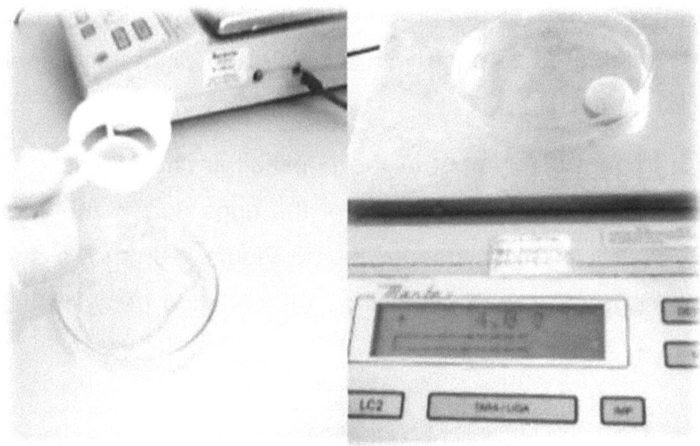

Figure 17. Weighing of yolk and albumen separately.

4.2.5 Pigmentation of egg shells

In addition to their small size, quail eggs show pigmentations on the shell as a classic characteristic of identification of the species from which they originate. Some authors even characterize the quantity and size of pigments (spots) with reproductive and egg quality factors, as well as the brightness or opacity of the shell. In the case of japonica quails, eggs that are shiny and with several pigments are characterized as being of better quality (Albino and Barreto, 2003).

The pigmentation of the eggshell of quail eggs denotes the time of permanence of the eggs in the oviduct. Thus, the more pigmented eggs are those that remained for a longer period, while the less pigmented eggs remained for a shorter time, the latter being considered immature eggs, often characteristic of highly productive birds.

Albino and Barreto (2003) described that opaque eggs with bluish shells are characteristic of birds that took 36 to 48 hours to lay, i.e. those that remained for a long period of time in the oviduct. These characteristics end up reducing the porosity, with consequent reduction of shell permeability. On the other hand, eggs with a reduced number of pores in the shell end up losing less weight during storage compared to eggs with a shiny and pigmented shell (Albino and Barreto, 2003).

The shape, size and amount of pigment in the eggshell is characteristic of each quail, being possible to identify the origin of the eggs due to the very similar pigmentation characteristics among them. Figure 18 shows the similarity between the eggs of a quail, as well as the variation between the eggs of other birds.

Figure 18. Detail for pigmentation similarity between eggs of the same quail.

Eggs are usually darkly pigmented, with brown or black spots, and may vary in shades of yellow, brown, bluish or greenish. The shell color is the result of the deposition of some minerals, such as calcium, copper, sodium, iron and pigments biliverdin and porphyrin (Albino and Barreto, 2003). Besides being related to quality, the shell pigmentation of quail eggs is also an important characteristic for their commercialization, being considered a selection factor by the consumer.

4.2.6 Sensory analysis of eggs

According to Seibel et al. (2010) egg quality is no longer evaluated only by the condition of the shell, white or yolk. In recent years, it has also come to be measured through physical properties and nutritional or chemical modifications that can affect the eggs.

Sensory analysis is a tool used to verify the sensory characteristics of a given food, where some attributes such as odor, texture, flavor, hardness, among others, can be observed.

To characterize a food, a descriptive sensory analysis based on predefined terminologies for obtaining sensory perceptions is considered as a primary step (Seibel et al., 2010).

According to Noronha (2003), sensory analysis allows differentiating and characterizing sensory attributes of products, as well as determining whether the differences found are perceptible or accepted by the consumer.

The taste of a food, for example, is considered a decisive attribute in the choice and acceptance of a food, being a response that is integrated to the sensation of taste and aroma of the product. Taste is an attribute that allows the detection of volatile compounds in foods, including the presence of sugars, some salts and acids that determine the so-called four basic tastes: sweet, salty, bitter and sour. Aroma, in turn, denotes the presence of various volatile substances with different physical and chemical properties, and is considered a more complex attribute to evaluate (Thomazini, Franco, 2000).

The sensory evaluations are carried out in individual booths (Figure 19), with the absence of noise and odors, at previously established times, excluding one hour before and two hours after lunch, according to the method cited in Moraes (1985). The composition of the panel of tasters consists of at least 10 trained people, who can be of both sexes and of different ages.

The selection of candidates to compose the trained panel can be followed through the adaptation of the methodology cited by Seibel et al. (2010). When initially the selection of a group of tasters must be done, through discriminatory tests, in which the tasters are instructed to recognize the attributes flavor, aroma and hardness perceived in the samples served.

After the training period, the tasters carry out three evaluation sessions. In each of these, the eggs are cooked in boiling water for seven minutes and after reaching room temperature, they are peeled, if possible with the aid of a quail egg peeler, thus maintaining a pattern of appearance and integrity of the eggs.

Samples can be served in disposable plastic cups, coded with random three-digit numbers, accompanied by cookies, mineral water at room temperature (for residual flavor removal) and coffee beans (for odor removal) between samples.

In all sessions the tasters should be distributed in individual booths and oriented to evaluate one sample at a time from left to right. To avoid confusion in the answers, a red light should be turned on in the testers' booths to prevent visual discrimination of the samples by color.

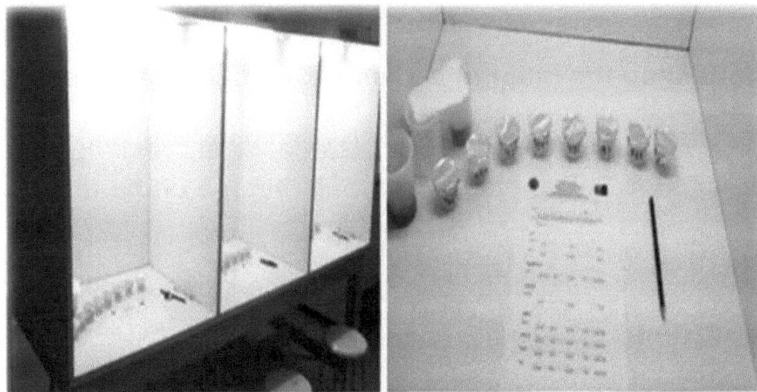

Figure 19. Individual booths for sensory analysis of foodstuffs.

In the first session, the evaluation can be done through the sensory attributes of the eggs on a structured 11 cm scale. The tasters in their individual booths receive a card to evaluate the attributes of color (intensity of yellow color and brightness), odor (egg characteristic), flavor (acid, residual, rancid) through 11 cm structured scales, anchored at the ends on the left by the term "little or absent" and on the right by the term "strong or very strong".

Tasters should be instructed to indicate with a vertical line under the scale line, the point that best represents the perceived intensity of each characteristic (Stone and Sidel, 1998).

In the second evaluation session, the discriminatory test "duo-trio test" can be performed. This is a procedure to detect the sensory difference between a sample and a standard. In this case, the judges are presented simultaneously with the standard and two coded samples. In this evaluation, the taster must identify the sample equal to the standard (IAL, 2008).

For the third evaluation session, the triangular test can be used. This procedure is similar to the previous test, but in this case three coded samples are presented simultaneously, two being the same and one different. The taster must identify the different sample. The probability of success is $P = 1/3$. In the same way as the test of the previous session, the interpretation of the result is based on the total number of judgments versus the number of correct judgments.

4.3 Characteristics affecting the internal and external quality of eggs

The age of the bird directly influences the quality, composition and size characteristics of the eggs, since with the increase in the age of the matrix, changes occur such as, for example, a reduction in the laying rate, an increase in egg size and alterations in the constitution of the yolk and albumen (Rocha et al. 2008).

Knowing the effect of quail age on egg quality is important, as it is known that deterioration of albumen and shell quality are causes of reduced incubation performance (Nowaczewski et al., 2010).

5. Quail management

5.1 Sexing

The development of industrial poultry farming is largely due to the use of adequate management techniques. Among the management techniques, sexing by plumage color stands out as a procedure that brings profit to the activity and has been widely used in the last decades in production farms.

The method consists in the separation of male and female chicks into differentiated flocks based on the visualization of feather coloration shortly after hatching. In japonica quails, females have brownish coloration and males have darker shades approaching black. The ideal time to identify the sex of quails is between 10 and 24 hours after hatching, when, in a very simplified way, the female is darker and the male is redder.

With the separation of flocks into male and female chicks, the producer obtains gains from the possibility of producing uniform birds with average weight adequate for each market demand. Taking into account that the nutritional needs of both sexes are different, as well as the management.

Sexing, therefore, is a technique used to differentiate newly hatched chicks into males and females.

In addition to the two forms of sexing already mentioned, it is still possible to differentiate between male and female quail by visualization of the cloaca. However, sexing by cloaca is a method not available to many and only mastered by a few agribusiness technicians.

At the age of 17 to 20 days it is possible, in the case of European quails (*Coturnix coturnix coturnix coturnix*), to separate males and females by visualization of the feathers around the neck and breast of the birds (Figure 20). In this case, the males have brown or dark beige feathers around the neck and breast and the females have lighter (white) breast feathers with dark spots and no dark coloration on the neck.

Figure 20. Differentiation of male (left) and female (right) European quail (*Coturnix coturnix coturnix coturnix*).

5.2 Reproductive management

The mating of quails can be done throughout the year and occurs in the ratio of 3:1, that is, three females for one male, achieving a good percentage of fertilization of the eggs.

An important aspect in relation to the mating of quails is the high sensitivity of these birds to inbreeding, with generation of defective animals at birth. Therefore, the recommendation is to avoid mating between close generations as much as possible. In this sense, in order to avoid inbreeding-related events, such as yield reductions, producers are constantly seeking to incorporate new lines to improve their breeding.

At about 45 days of age, the quails are ready for mating and subsequent laying of fertilized eggs. This mating can be done daily by taking the male to the cage of the respective females.

Cage batteries should be located in well ventilated sheds or aviaries with an ideal comfort temperature for the birds, since excessive heat lowers the fertility of the males.

5.3 Chick handling

After hatching, when about 95% of the quail chicks are "dry", they should be placed in an environment with a heating temperature of about 38 to 39°C (100 to 104°F). The temperature should be gradually decreased until the birds no longer need artificial heat. From the third day of life, the temperature should be reduced daily by 1°C until the temperature is equal to that of the environment.

Water should be offered at will, taking care to wet the beaks of some quail so that the others, through observation, learn where to drink. Manual drinkers should be washed and the water should be changed at least three times a day, being more recommended the automatic drinkers that supply the chicks with fresh and clean water in abundance.

The floor where the chicks are to be housed should be lined with paper or cardboard on wood shavings during the first three days of life. The initial feed can be distributed as desired on this lined surface. Subsequently, the feed will be offered in tray type feeders.

The birds are raised in this environment until they can be transferred to poultry houses or batteries for breeders, at which time the selection of birds for breeding, laying or fattening takes place.

5.4 Rearing management

The rearing period is between 16 and 45 days of age. During this period, the birds continue to receive feed and water ad libitum. At 30 to 35 days of age, the females can be housed in cages and start receiving starter feed.

5.5 Layer management

The amount of feed per bird/day should be 30 to 35 grams, and water should always be supplied ad libitum. For a high laying rate, the environment of the quails in production should be well illuminated for 17 hours, preferably by natural light and with the help of 15 watt lamps, one for each 5.00 m_2 of house.

5.6 Egg handling

Eggs to be marketed should preferably be collected twice a day. The first collection should be done in the morning and the other in the late afternoon. The eggs should be packed in suitable containers, kept refrigerated so that their nutritional qualities are preserved.

It is important to emphasize that eggs intended for human consumption should not be fertilized, that is, they should not come from productions with the presence of males. This precautionary measure serves to avoid possible surprises for the consumer when opening the eggs at home, because it is once fertilized and depending on the temperature and time it is possible that the embryo develops.

In this sense, Brazilian legislation does not allow the commercialization of fertilized eggs through a federal law. Such requirements and regulations can be found and consulted at the federal, state or municipal inspection services. However, despite the existence of regulations, it is possible to find and purchase these products in informal markets. Thus, eggs intended for human consumption in a regulated manner come from farms that raise only female quail and, therefore, their eggs are not fertilized.

5.7 Hatchery management

Eggs destined for incubation are those resulting from mating between males and females in a 3:1 ratio, i.e. three females for each male. Within 48-72 hours after mating, the eggs may already be fertilized. However, to ensure that all the eggs are fertile, it is usually about 10-14 days after mating before the first collection for incubation.

It is advisable that quail eggs should not be stored for more than eight days, since after this period hatching rates tend to decrease drastically. After collection, they should be stored with the air chamber upwards, i.e. with the pointed side downwards and in a cool place.

For egg incubation, the natural method can be used, however, as female quail do not normally have the characteristic of setting eggs, miniature laying hen females can be used, which do this very well.

The artificial incubation method uses everything from home incubators (manufactured or adapted by the producer) to industrial incubators of various sizes. Figure 21 illustrates an artificial incubator with a capacity to incubate up to 2,000 quail eggs.

Eggs destined for hatching must be fertilized. Therefore, farms producing fertile eggs have female and male quails reared together in their stock.

Eggs for hatching should be stored in the sheds for a short period of time and sent to the hatchery at least twice a week.

Figure 21. Artificial incubator with capacity for 2 thousand quail eggs.

The Ministry of Livestock Agriculture and Supply (MAPA) in Brazil through Normative Instruction (IN) n° 05 of 02/14/2017 - Art.64 - Sole Paragraph determines that in the production of quail eggs the candling stage is dispensed with.

Candling is a tool used for the internal inspection of eggs intended for incubation. Candling allows to observe if the eggs are fertile and if the embryos are alive. This process makes it possible to discard infertile eggs or eggs with dead embryos. However, this method is more commonly used for grading and observation of hen eggs (Figure 22).

Figure 22. Candling process for grading hen eggs (not performed on quail eggs).

After grading, the eggs can be placed in the hatchery trays. As a biosecurity measure, the eggs go through a sanitization process to maintain their integrity with respect to the presence of pathogens. Among the biosecurity measures that can be adopted is dry sanitization or fumigation (a solution of formalin and potassium permanganate). This operation should not be performed when the eggs have been stored for more than 12 days.

Fertile eggs can be stored in specific rooms, called egg rooms, which are intended for the storage of hatching eggs until they are placed in the incubator. It is important that the environments where the eggs are stored (shed, transport and egg room) have similar conditions to avoid sudden changes in temperature and humidity that cause condensation (transpiration), cooling or heating of the eggs.

Egg storage has some important effects:

Birth decreases with prolonged storage time - the effect increases as storage time is prolonged, (after 6 days, a reduction of 0.5 to 1.5% daily occurs) with a progressive increase as the days pass;

Hatch rate and chick quality are seriously compromised in eggs stored for 14 or more days.

Gas exchange through the pores of the eggshell occurs during storage. Carbon dioxide is dispersed out of the egg and its concentration decreases. Eggs also lose moisture during storage. The loss of carbon dioxide and moisture during storage contributes to reduced hatchability and chick quality. In this regard, storage conditions should be such that these losses are minimized.

Eggs can be stored in closed or open boxes, being in most cases placed in incubation trays on trolleys. The temperature should be around 18°C and the relative humidity around 75%. In order to avoid condensation and consequently fungal growth, it is important that during storage the eggs are thoroughly cooled and dried before incubation.

The incubation period for quail eggs is 17 days, which is three days less than for chicken eggs. The factors that determine incubation time are: temperature, humidity, and eggshell quality.

The ideal temperature in forced ventilation incubators should be constant and remain around 37.7°C while the humidity should be 60%. If the eggs are transferred to the hatcher (about three days before hatching), the humidity should be around 70 to 75% and the temperature should be close to 37°C.

The hatcher differs from the setter for two main reasons: in the setter the eggs are placed on trays with individual spacing and with the thinner end of the eggs facing downwards. The second reason is related to the turning of the eggs in the setter, where the trays are moved horizontally at an angle of up to 45°C (Figure 23).

Egg turning aims to prevent the embryo from adhering to the eggshell membrane, especially during the first week of incubation. In addition, turning also aids in the development of the embryonic membranes. As the embryo develops and increases its capacity to produce heat, constant turning helps air circulation and aids temperature reduction.

Figure 22. Diagram of egg turning in the incubator trays at 45°C angle.

Figure 24 shows a domestic artificial incubator in which plastic trays used for marketing eggs for the hatching of chicks can be adapted at the bottom.

In industrial incubation, about three days before the expected hatching date, the eggs are transferred to the hatching chambers. The eggs are arranged in a single compartment tray that will allow the chicks to break through the eggshell more easily and with more space.

Figure 23. Domestic egg incubator with trays adapted for hatching.

The transfer of the eggs to the incubation chamber is done for two main reasons: 1) the eggs are left on their side (lying down) to facilitate the free movement of the hatching chick; 2) it facilitates hygiene during hatching, when a large amount of feathering is produced that can contaminate the incubator.

When eggs are transferred too early or too late, the embryo is exposed to less favorable conditions, thus decreasing hatchability. Therefore, the transfer should be done carefully and quickly avoiding cooling of the eggs, which can result in delayed hatching or death of the embryo.

It is important to note that at this stage the eggshell is more fragile, as the embryo removes calcium from the shell for the formation of its skeleton. Therefore, extreme care must be taken during the transfer to avoid egg breakage. Another detail to consider is that the trays must be clean and dry at the time of transfer, since wet trays cool the eggs when the water evaporates. In addition, improper handling of the eggs during this phase can cause embryo breakage and bleeding.

The success of a hatchery is measured by the number of quality chicks produced. On the other hand, success in chick production also depends on the quality of the eggs at farm level.

Fertility, for example, is a factor completely influenced by on-farm management, considering that the hatchery cannot improve egg fertility. Among the problems that can arise during the incubation process we have:

a)   Unfertilized eggs - few males, old males, sterile on farm;

b)   Early embryo death - incubator temperature too high or too low, low vitamin content in feed or lack of fumigation of eggs;

c)   Late embryo death - improper temperatures, inadequate turning and insufficient ventilation;

d) Chopped eggs with dead chicks - low humidity, medium-low temperature or over-temperature;

e) Early hatching - high temperature in the incubator;

f) Late hatching - very low incubator temperature;

g) Chicks stuck to the shell - low humidity during hatching;

h) Malformed chicks - lethal factors or other hereditary causes and irregular temperature.

To avoid problems with egg incubation some care should be taken during the handling of hatching eggs, among them:

a) Avoid thermal shock to the embryo;

b) Provide good air circulation around the eggs;

c) Always handle the eggs carefully, preventing cracks;

d) Discard eggs unsuitable for hatching by taking strict care during egg selection;

e) Keep the pointed end of the egg down;

f) Store eggs in a separate chamber with controlled temperature and humidity;

g) Ensure proper and complete hygiene.

h) During the incubation operation it is important to maintain correct temperature and humidity.

i) Allow adequate gas exchange with frequent turning of eggs.

Chicks are ready to be removed from the hatcher when most of them are dry (5% with still wet necks are acceptable). A common mistake is to leave chicks in the hatcher longer than necessary, which causes chick dehydration. Chick dehydration can be the result of an error in setting the setter time in relation to the age of the eggs. If the eggs are "green" at hatch, it is necessary to check the possibility of cooling during the incubation process, which reduces the degree of development. While the chicks are being removed from the hatcher, they should be graded according to their quality (top quality or discarded) and only then released to their destination.

5.8 Factors interfering with the incubation of fertile eggs

5.8.1 Nutritional management

In addition to adequate temperature, humidity, periodic turning, ventilation and positioning, the embryo needs all the essential nutrients (proteins, carbohydrates, fats, minerals and vitamins) for development during the incubation period.

Hatchability and chick quality at hatch can be influenced by a number of factors, such as the age of the matrix, the nutrients supplied by the feed and the storage time of the eggs prior to incubation. Intrinsic factors are also of extreme importance on the success of egg incubation, such as humidity, temperature of the storage room, incubator and hatcher.

5.8.2 Breeder's age

Hatching performance, chick weight and chick quality depend on several factors including, among others, the age of the breeder which, in turn, influences egg weight (Rocha et al., 2008).

Eggs from young breeders have thicker shells and denser albumen, which reduces moisture losses and gaseous changes (Brake et al., 1997). These factors, together with the low capacity of young birds to transfer lipids to the egg yolk, end up compromising embryo development, embryonic viability and, consequently, egg hatching (Fasenko, 2003).

Older breeders produce heavier eggs and therefore end up having lower hatchability, as the embryos developed in larger eggs are less resistant to the metabolic heat produced at the end of the incubation period (Lourens et al., 2006).

However, older breeders produce eggs with larger yolks. Therefore, the eggs will provide more nutrients to the chick at hatch (Rocha et al., 2008). In this sense, larger eggs exert a positive influence on weight gain and consumption of European quail at 21 and 42 days of age.

5.8.3 Feed nutrients

Among the most important factors for the success of a breeding operation, whether for meat or egg production, is the quality of the feed.

For this reason, the ingredients used in the manufacture of feed must be of guaranteed origin and quality. The ingredients must have high digestibility rates for a better utilization of nutrients.

The digestibility of nutrients and their energy value are influenced by the rapid transit time of digesta through the intestine (1 to 1.5 hours in quails versus 3 to 5 hours in chickens), and quails are able to better utilize the energy from the feed fiber due to the greater relative size of the cecum (Sakamoto et al., 2006). On the other hand, the use of lipids reduces the transit speed of digesta through the gastrointestinal tract, making it possible to improve the absorption of all dietary ingredients (Baião and Lara, 2005).

Nutritionists recommend a minimum level of 2% linoleic acid in the poultry diet at the beginning of production. Scragg et al. (1987) verified that levels of up to 2.7% linoleic acid in the diet provided an increase in egg weight. However, other studies have already shown that levels above 1.2% and 2.4% of linoleic acid (Brake et al., 1989) do not influence egg weight.

It takes 2% of linoleic acid in the diet to provide egg weight gain (Menge, 1968) and this effect is more significant in younger birds (Whitehead et al., 1991). In addition, the use of oils rich in polyunsaturated fatty acids in the diet improves hatching and performance of the birds (Baião and Lucio, 2005).

The albumen and yolk are responsible for the supply of nutrients to the embryo (Santos et al., 2009), with the yolk sac being the main reservoir of nutrients in the first 24 hours of chick life (Vieira and Moran Jr., 1999).

According to Ribeiro et al. (2008), due to better hatching development, heavier chicks may have more well-developed carcasses and smaller yolk sacs, or less developed carcasses and larger yolk sacs, which increases survival before starting exogenous feeding.

With respect to quail performance, the divergent results found in the literature can be explained by the fitness, genetics or line studied and also by the nutritional levels of the diets.

The inclusion of certain types of oil in the diet, such as canola oil, promotes the incorporation of highly unsaturated fatty acids in the egg yolk, which consequently increases its oxidative potential. In this sense, Paton et al. (2002) verified that the micromineral selenium has beneficial effects on embryo development.

In reproduction, selenium is one of the main minerals that act as antioxidants improving male fertility and thus improving reproduction (Alonso et al. 1997). Antioxidants act by protecting the mitochondrial membrane present in spermatozoa which are responsible for motility, fluidity and increased membrane flexibility. However, to perform these functions, mitochondria require high levels of polyunsaturated fatty acids. However, these high levels end up leaving the cell with increased free radical production and vulnerability to oxidation (Surai, 2002).

Among the factors related to feed ingredients, individual variations, moisture, frequency, time of feed delivery (Murakami and Furlan, 2002), feed composition, physical appearance and amount ingested are determining factors that are directly related to feed efficiency of animals (Leandro et al., 2001).

6. Nutrition and food

One of the main functions of food is to provide energy to the organism for the performance of all animal functions. However, it is important to know their chemical and physical nature, since they influence both the consumption of dry matter and the digestion and utilization of nutrients.

Costs with quail feed tend to be higher compared to those of laying hens and chickens, considering that quail feed contains a higher concentration of protein (Silva et al., 2012).

On the other hand, with the advance of new research, new data related to quail nutrition have emerged in the national and international literature (Silva et al., 2012). With the results of this research, it is increasingly possible to formulate diets with lower cost and better economic return.

Although the nutritional requirements of quail are not the same as those of laying hens and broilers, they use energy from corn and soybean bran in a similar way. However, it is not advisable to feed quail with chicken and laying hen feed, as quail require less calcium and more amino acids (Silva et al., 2012).

In terms of genetic improvement, while broiler (European) quails have been selected for high rate of weight gain (Aggrey et al., 2003), laying (Japanese) quails have been bred for high egg production, in an attempt to produce more nutritious, better quality products with lower cholesterol content (Minvielle and Oguz, 2002).

Fattening quails show a faster growth in relation to laying quails, reaching their growth peak around 27 days of age, where there is probably a greater deposition of protein and water in the carcass. After this point, growth decreases and there is a greater deposition of fat in the viscera with retention of nutrients in the ovary, in the case of females, and an increase in dietary requirements (Silva et al., 2011).

At five weeks of age, fattening quails increase about 25 times their initial weight, and females can weigh 10% more than males between the sixth and eighth weeks of age (Silva et al., 2006).

Since feeds formulated from the nutritional recommendations for broilers or laying hens do not have everything that quails need for production, it is important to know the nutritional requirements for the optimization of the productive capacity of these birds.

For the formulation of quail diets, nutritional requirement tables from other countries, such as those of the NRC (1994) and INRA (1999), are normally used. However, these tables may present extrapolated values of metabolizable energy and nutrient requirements for broilers, layers and turkeys (Barreto et al., 2006). Therefore, they may not correctly meet the requirements of these birds.

In addition, in recent years this bird has been undergoing a constant process of genetic improvement for better meat or egg production. Therefore, it is important to remember that the tabulated values should be taken with caution and should be updated by nutritionists and producers (Villela, 1998).

On the other hand, the nutritional requirements of quail can vary, among other factors, with age, sex, environment, energy levels and amino acids. However, requirements are usually estimated according to the amount of nutrients required to perform basic body functions and productive functions most efficiently (Garcia et al., 2006).

Dietary energy, in general, comes from the use of proteins, carbohydrates and lipids, being lipids the best sources of energy to be used by animals, because besides supplying energy with low caloric increment, they are sources of essential fatty acids for the maintenance of the structure and function of the cellular membrane.

The nutritional properties of foods are directly related to the quantity and quality of their constituents, which can be divided into primary and secondary. Primary constituents include proteins, carbohydrates, lipids, vitamins, minerals and water. Secondary constituents include enzymes, organic acids, pigments and others. Primary and secondary substances are responsible for the nutritional and sensory characteristics of food and act in a differentiated way. Therefore, birds need to receive the right proportions and quantities of nutrients to be able to effectively perform their productive and reproductive functions.

6.1 Proteins and amino acids

Proteins are made up of amino acids and represent the main structures of most of the body structure of birds. They also play a role in tissue synthesis, body metabolism, egg size and number.

The so-called essential amino acids are those that are not produced by the animal organism or are not produced in sufficient quantities. Thus, only nine of the 20 amino acids required by quails are considered essential (D'Mello, 2003).

As for amino acids, the most used are methionine, lysine and threonine, because they are, respectively, the first, second and third limiting amino acids in poultry feed (chickens and layers), being essential for development, growth and egg production.

Methionine is the first limiting amino acid in corn and soy bran based rations for quail (Mandal et al., 2005). Being the first amino acid of the polypeptide chain of proteins, acting in the transport and donation of methyl groups for choline synthesis from ethanolamine, it donates sulfur to serine, which results in the synthesis of cysteine, and can be an indicator of the nutritional status of the animal in vitamin B12 (D'Mello, 2003).

Threonine is considered the third limiting amino acid in corn and soy bran-based diets for chickens and turkeys, but it is the second, after methionine, in quail diets (Mandal et al., 2006). Threonine has no intermediate precursors, and the D isomer cannot be converted in the organism to the L isomer (D'Mello, 2003), so it is necessary that the diet contains 100% of the poultry's needs. Threonine is still involved in the synthesis and secretion of mucin, amylase and intestinal mucosa growth.

The amino acid profiles of the protein sources used in poultry diets do not normally meet the metabolic needs for maintenance, tissue growth and products (Bertechini, 2006). Since poultry do not store protein, the daily supply of adequate amounts of industrial amino acids becomes necessary to meet poultry requirements without excesses or deficiencies, without impairing animal growth or increasing feed costs (Macari et al., 2002).

However, the decrease in body protein synthesis and protein deposition in the egg may also be a result of decreased digestibility (fraction of feed consumed and not recovered in the excreta) which, in turn, alters the efficiency of utilization and bioavailability of most amino acids (Silva et al., 2006).

In this sense, the use of nutritional strategies to improve the composition and quality of animal products destined to feed the population is a link between animal production, food technology and human nutrition (Barreto et al. al., 2006). Among the protein sources, soybean bran, due to its amino acid balance, is the main ingredient used in diets (Albino and Barreto, 2003).

6.1.1 Ideal protein

Protein is one of the main nutrients used in the formulation of poultry diets. It represents a considerable cost in poultry production since it must be added in large quantities and is not cheap. However, it is essential, since it is responsible for protein deposition in the carcass, influencing weight gain and feed conversion of the animals.

Initially, poultry diet formulations were based on crude protein levels. Later, formulations were adjusted according to total amino acid levels to meet the protein requirements of the animals. However, the formulation based on total amino acids tends not to reach the ideal protein levels mainly due to variability of amino acid digestibility in different ingredients. In this case, it is possible to supply levels of some amino acids below or above the animal's requirements. In the latter case resulting in wastage with consequent increase in production costs with the feed (due to increased protein), in addition to the increased environmental impact due to increased nitrogen excretion from animal excreta.

In this sense, as each feed has its own digestibility value, protein levels are best achieved when the levels of digestible amino acids are considered in the formulation of diets. Feed composition tables provide the digestible amino acid values necessary for the formulation of diets, allowing nutritionists to reach with the formulation the real protein needs of poultry (Rostagno et al., 2011).

The concept of ideal protein arose from the need to adjust diets to the real needs of the animals avoiding as much as possible the waste of ingredients in the formulations with consequent improvement of feed utilization by the birds, in addition to improving their performance.

Ideal protein can be defined as the exact balance of essential amino acids with the supply of non-essential amino acids. Thus, ideal protein based formulations must be able to meet the maintenance and production needs of the animals without underestimating or overestimating any amino acid while still allowing maximum protein deposition in the carcass. In this sense, it is possible to formulate a diet with greater economy and better utilization of amino acids reducing the environmental impact due to lower protein nitrogen excretion.

In the concept of ideal protein, a standard amino acid is used, from which it is possible to calculate the necessary quantity of the other amino acids.

The standard amino acid chosen was Lysine because it is used more for protein synthesis, because it has a greater number of published works and because its analysis is less expensive than that of methionine and cysteine (Jordao Filho et al., 2006).

The amount of reference amino acids should be calculated considering the ideal percentages or proportions according to the birds' needs, respecting their physiological state, age, as well as anti-nutritional and genetic factors (line, sex, body condition), in addition to environmental conditions (Baker et al., 2002).

The addition of synthetic amino acids in the formulations allows lower inclusions of ingredients containing high protein indexes, such as soy bran, or rich in energy (oils and fats), reducing the cost of the diets.

Therefore, the concept of ideal protein allows the nutritional requirements of the animals to be met, with better utilization of ingredients and reduction of feed costs.

6.2 Carbohydrates

Carbohydrates are aldose (polyhydroxyalkoxy aldehyde) or ketoses (polyhydroxyketones) and can be divided into monosaccharides, oligosaccharides or polysaccharides. These substances are found in natural foods and constitute the main source of energy available in poultry feed.

Starch (polysaccharide) is a source of energy for animals and plants and can be found in roots, seeds and tubers. It consists of two types of glucose polymers, amylose and amylopectin (Figure 25), which vary in species and degree of maturation.

Amylose is formed by a linear chain of D-glucose units, linked together by α-1,4 glycosidic bonds, and can contain from 350 to 1000 glucose units in its structure. It has a helical structure, α-helix, formed by hydrogen bridges between the hydroxyl radicals of the glucose molecules.

Amylopectin consists of linear chains of 20 to 25 D-glucose units linked at α1,4. These chains are connected to each other through α1,6 glycosidic bonds, forming the branches, which occur between every 24 and 30 glucose residues. Amylopectin contains 10 to 500 thousand glucose units and has a spherical structure.

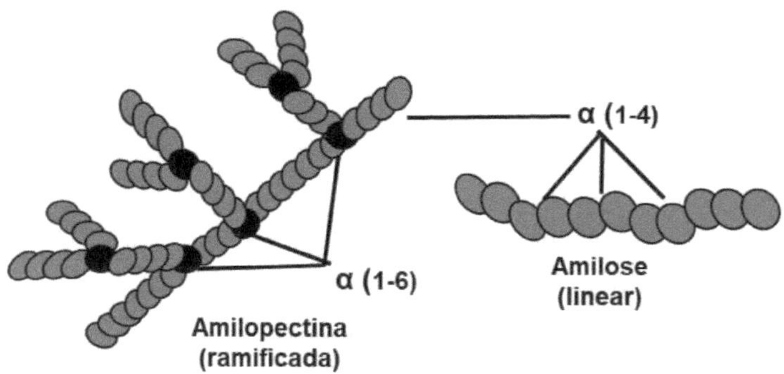

Figure 24. Diagram of the chemical structure of amylopectin and amylose.

The end products of carbohydrate digestion are simple sugars that are metabolized to produce water, CO_2 and energy. Starch, cellulose and glycogen are the main glucose polymers polysaccharides of importance in poultry nutrition.

Starch stands out as the main digestible polysaccharide of plants, present in large quantities in cereal grains. Cellulose is a glucose polymer with beta 1-4 bonds. The digestibility of cellulose for poultry is generally very limited, however, it plays an important role in controlling the rate of passage of digesta through the digestive tract of birds. Glycogen, in turn, is the reserve form of carbohydrates in the animal organism serving as a source of energy for direct utilization by birds.

Among the sources of carbohydrates used in the manufacture of feed, corn, sorghum and wheat bran stand out. Alternative sources are also recommended in cases of attempts to reduce feed costs (Albino and Barreto, 2003).

6.3 Lipids

Fatty acids are the main components of the lipid structure (Gomes and Tirapegui, 2000), which in turn are the hydrolysis products of triglycerides and are present in animal and vegetable fats in even number of carbons, due to the biosynthesis of two carbon units (Bertechini, 2006).

Fatty acids are necessary for various physiological functions. Monogastrics are unable to synthesize linoleic (ω-6) and linolenic (ω-3) fatty acids, so they are considered essential and therefore must be supplied through the diet (Dolz, 1996).

According to Butolo, (2002) fatty acids are classified according to their chain length and are divided into: short chain fatty acids (with less than eight carbons); medium chain fatty acids (with eight to 11 carbons); intermediate chain fatty acids (with 12 to 15 carbons) and long chain fatty acids (equal to or greater than 16 carbons).

Depending on the presence or absence of double bonds, fatty acids can be defined as saturated (those without double bonds), monounsaturated (those containing one double bond) and polyunsaturated (when two or more double bonds are present).

Unsaturated fatty acids are characterized by having 16 or more carbon atoms in their chemical structure, two or more double bonds that are differentiated into several series or families, such as omega 9 (ω-9), omega 6 (ω-6), omega 3 (ω-3), among others. The most important families are omega 9 (C18:1), omega 6 (C18:2), derived from linoleic acid (LA), and omega 3 (C18:3), derived from α-linolenic acid (LNA). Depending on the position of the first double bond, from the methyl group, the names of the series are derived on the sixth or third carbon atom, respectively (Briz, 1997).

Lipids participate in the regulation of animal metabolism, being part of the structure of prostaglandins and steroid hormones, and their primary function as a nutrient is the production and storage of energy, but they are also important sources of essential fatty acids (Bernardino, 2009).

lipids also play an important role in the production and regulation of eicosanoids. Eicosanoids are derived from linoleic acid (ω-6) and arachidonic acid (AA) and will be precursors of prostaglandin E2 (PGE2). However, the desaturase enzyme, which has a function of decreasing arachidonic acid production, is inhibited by linolenic acid (ω-3), which implies a decrease in PGE2 (Hirayama et al., 2006). Thus, it is supposed that the high consumption of foods rich in linoleic acid added to the low consumption of linolenic acid could result in disturbances in the production of eicosanoids.

Fatty acids constitute the basic units of lipids and their determination is fundamental for the knowledge of the quality of oils and fats, for the verification of their nutritional value and the effects of food processing.

Lipids as a source of energy in diets produce 2.25 times more calories compared to carbohydrates. According to the NRC (1994), the use of fats in feed formulations provides a beneficial effect to poultry by improving the efficiency of utilization of energy consumed, due to the lower caloric increase of lipid metabolism, palatability, feed conversion and vitamin absorption. It also reduces pulverulence and nutrient losses (Nunes, 1998).

The use of oils in poultry diets is used when it is desired to increase the energy level, improve feed conversion, increase the absorption of fat-soluble vitamins and the efficiency of energy consumption. For this purpose, different lipid sources can be used, such as soybean oil, canola oil, fish oil or a mixture of these (Baiâo and Lara, 2005).

Energy influences the performance and carcass yield of poultry. In addition, it affects the cost of the diet due to the large volume of ingredients in its composition (Albuquerque et al., 2003).

6.3.1 Essential fatty acids

Essential fatty acids are not synthesized by animals. Therefore, they must be supplied through the diet because they are essential for the development of animals. Essential fatty acids are represented by fatty acids of the omega-3 (linolenic acid) and omega-6 (linoleic acid) families. The latter acts on enzymatic functions, fluidity and receptors of animal cell membranes.

Arachidonic acid is formed by the conversion of linoleic acid into a longer chain polyunsaturated fatty acid, and can be converted by the animal organism into other important long-chain fatty acids that act as biological mediators (Butolo, 2002).

Although linoleic and arachidonic fatty acids are considered essential to the animal organism, they are produced in the liver by the synthesis of arachidonic acid from linoleic acid in the presence of vitamin B6 (Bertechini, 2006). Therefore, only linoleic acid (C18:2) is considered negatively.

Linolenic acid (ω-3) belongs to the omega-3 (C18:3) family of polyunsaturated fatty acids and, in turn, can give rise to a number of omega-3 fatty acid components, prominent among them being eicosapentaenoic acid (C 20:5), which is referred to as EPA (Butolo, 2002).

Monounsaturated fatty acids, such as, for example, oleic fatty acid, are considered fundamental in the reduction of LDL-cholesterol oxidation (Angelis, 2001). Therefore, they become important in the human diet, as they have been linked to the prevention of dyslipidemias.

Dyslipidemias are designated as responsible for the development of coronary and cardiovascular diseases. In this sense, oleic fatty acid has been related to a decrease in serum cholesterol levels, a decrease in low-density lipoproteins and an increase in high-density lipoproteins in individuals with hypercholesterolemia, as well as a reduction in blood glucose levels in diabetic patients (Soares and Ito, 2000).

The fatty acids that make up oils are mostly saturated (SFA) and unsaturated (UFA) fatty acids, which in turn can be monounsaturated (MUFA) or polyunsaturated (PUFA), the most important being those of the omega-9, omega-6 and omega-3 series (Valenzuela and Sanhueza, 2009). Fatty acids are structural constituents of cell membranes, fulfilling energetic functions and metabolic reserves, in addition to participating in the formation of hormones and bile salts (Valenzuela and Nieto, 2003).

Through dietary manipulation of poultry, it has been possible to improve the nutritional quality of eggs and reduce the concentration of saturated fatty acids in the yolk (Carvalho et al., 2009), thus promoting healthier foods acceptable to the consumer market.

However, vegetable oils, such as soybean oil or canola oil, have in their composition unsaturated fatty acids that are susceptible to deterioration in the feed and can reduce the nutritional value of the feed. In addition, it is known that poultry meat has a higher concentration of PUFA, therefore, it can be more susceptible to lipid spoilage (rancidity), and that is why it is important to use antioxidants in the diet (Pita et al, 2006).

On the other hand, considering the quality of human food, it is important to include foods, such as quail eggs, that contain these monounsaturated fatty acids, such as oleic fatty acid, as an alternative to the use of oils, allowing greater flexibility in the preparation of food.

6.4 Vitamins

Vitamins are organic compounds, necessary in minimal quantities, to promote growth, maintain life and the reproductive capacity of animals. Vitamins belong to different classes of chemical compounds, thus presenting diversity in their physical, chemical and biochemical properties when compared to carbohydrates, proteins, lipids, minerals and water. The main classification of vitamins is based on their solubility, being water-soluble (B and C complex) and fat-soluble (A, D, E, E, K).

Water-soluble vitamins act as coenzymes and are eliminated after metabolic reactions, and are therefore called maintenance vitamins. Fat-soluble vitamins are stored in the body and are called growth vitamins (Albino and Barreto, 2003).

The amount of vitamins present in foods is not constant and may vary according to the season of the year in which the plant was grown, the type of soil or the cooking process to which the food was subjected. Most vitamins are altered in the presence of heat, light, humidity and preservatives.

Although they are required in small amounts in the formulation of diets, vitamins when in insufficient or absent concentrations can cause deficiency symptoms (avitaminosis). Vitamins are essential for the correct functioning of the metabolism and for the physiological functions of birds, such as maintenance, growth and reproduction.

6.4.1 Vitamin E: antioxidant source

Vitamin E stands out as an important antioxidant element of the animal organism. Oxidative processes are disturbances in the state of equilibrium between the pro-oxidant and antioxidant systems of the animal organism that result in oxidative damage to lipids, proteins, carbohydrates and nucleic acids (Jordàn Jr. et al., 1998). According to Cominetti et al. (2011) lipid oxidation is the main chain reaction involving the formation of free radicals.

Lipid oxidation causes changes in the organoleptic characteristics of meat and eggs that reduce their quality. In addition to promoting the undesirable rancid taste, oxidation, when it affects the pigments of the muscle or yolk, promotes changes in color, appearance and acceptability of cold meat and eggs, respectively.

In order to avoid this deterioration in the quality characteristics of meat and eggs, supplementation with vitamins with antioxidant characteristics in the diet of the animals has emerged as a preventive treatment (Morrissey et al.,

1994).

In this sense, oxidation processes can be reduced with the use of antioxidants in diets, whether natural or synthetic. Examples of natural antioxidants are organic selenium and, in the case of vitamins, vitamin E, which, in turn, can be transferred to the adipose fraction of meat or eggs, improving their oxidative stability. In addition, allied to the antioxidant effect, there is also the possibility of enriching products with this vitamin, promoting a nutritionally enriched food (Barroeta et al., 2002).

Vitamin E is able to neutralize free radicals from membranes that protect phospholipids from lipid oxidation reactions (Barroeta et al., 2002). The α-tocopherol is the active form of vitamin E with the highest biological activity frequently found in feed compounds for diet formulation, its effect being enhanced when in association with organic selenium through the enzyme glutathione peroxidase (Ziaei et al., 2013).

According to Surai (2002), there are three levels of antioxidant defense in the animal organism. At the first level occurs the prevention of the formation of free radicals by the action of antioxidant enzymes such as glutathione peroxidase, which contains the micromineral selenium in its constitution. A second level of defense prevents and restricts fat peroxidation through the sequestration of free radicals that promote a chain of fat peroxidation reactions. Vitamin E is the main and most effective scavenger of these free radicals. According to Surai (2002), there is also a third level of protection, which includes the repair of damaged molecules through lipolytic and proteolytic enzymes.

According to Mahan and Kim (1999) meat quality is directly related to the use of vitamin E in poultry diets. Vitamin E, in the form of alpha-tocopherol, can exert a color stabilizing effect by retarding the oxidation of myoglobin (Faustman et al., 1998). For this reason, the use of vitamins in the formulation of diets started to be used for the enrichment of animal food intended for human consumption and not only to avoid deficiencies (Surai, 2002).

Vitamin E has been supplemented in poultry diets because it has an antioxidant function in the biological system (Colnago et al., 1984, Rice and Kennedy, 1988, Finch and Turner, 1996). This effect on oxidation reduction is enhanced when in association with selenium through the enzyme glutathione peroxidase (Ziaei et al., 2013).

The association of these two antioxidants, vitamin E and selenium, supplemented in animal feed, have been shown to influence the maintenance of the antioxidant system (Jordâo et al., 1998; Boiago, 2006).

6.5 Minerals

Minerals are inorganic substances that perform various functions in the body and must be obtained through food. The excess or deficiency of one mineral interferes in the metabolism of another, being that the necessary quantities of each mineral vary from micrograms to grams per day.

Minerals represent 3 to 4% of the live weight of poultry and constitute an important part of the animal organism, for this reason they are considered essential elements for good nutrition.

Minerals can be classified into macro or microminerals depending on the organic needs of animals. Macro minerals are calcium, phosphorus, potassium, sodium, sulfur, chlorine and magnesium while micro minerals are iron, zinc, copper, iodine, manganese, cobalt and selenium.

6.5.1 Micromineral selenium: antioxidant source

Among the microminerals, selenium stands out as an important antioxidant element in the animal organism. After birth, animals need antioxidant protection, which may initially originate from the maternal diet or may be provided by natural antioxidants, such as carotenoids, ascorbic acid, vitamin E, antioxidant enzymes, such as catalase or glutathione peroxidase. Another way of providing this protection can be through the use of enzyme cofactors, such as the microminerals selenium, zinc, manganese and iron.

Selenium is an essential micromineral and, as an integral component of selenoproteins, it participates in physiological functions and biochemical processes of the organism.

Until the 1950s, selenium was studied mainly for its toxicity and not for its nutritional effect. However, since then, it has been recognized as an essential element, and its use in the diet of rats and birds is important in the prevention of hepatic, muscular and vascular lesions (Combs, 2001).

On the other hand, selenium, in its organic form, has been recognized and widely used by the poultry industry as an additive that allows improving oxidative stability and, consequently, improving the final product. With its use in poultry diets, it is possible to reduce the incidence of muscular dystrophy, improve fertility and hatching, improve shell quality and even strengthen the immune system (Payne et al., 2005).

Selenium is an essential part of a variety of selenoproteins, glutathione peroxidase (GSH-Px) being the best known. The main function of glutathione peroxidase is to combat oxidative stress by acting on hydroperoxides and lipoperoxides, preventing them from causing cell damage. The action of selenium in animal metabolism is also associated with the production of amino acids and proteins, which are efficient antioxidants (Moreira et al., 2001).

Selenium is an essential part of the embryonic development process by fighting free radicals and preventing the deterioration of the egg membrane. Selenium, when associated with vitamin E, promotes a better protection of membranes by preventing lipid peroxidation, being essential to avoid muscular dystrophy. In addition, it also protects against the toxicity of heavy metals such as cadmium and mercury (Watanabe et al., 1997). Thus, since eggs are rich in fatty acids, organic selenium acts to protect the embryo against oxidation reactions, improving its development through the action of glutathione peroxidase (Surai, 2000).

On the other hand. Pan et al. (2004) found that replacing inorganic selenium with organic selenium increased yolk and albumen weight. In most cases, a positive correlation is observed between hatching egg weight and chick size at hatch (Shanawany, 1987, Pinchasov, 1991, Wilson, 1991 and Rocha et al., 2008) representing about 62-76% of hatching egg weight. Therefore, having results indicating the increase in egg weight with the use of organic selenium and data proving the correlation between egg and chick weight, it is reasonable to infer that selenium improves the live weight of the progeny in selenium treatments.

Selenium can be consumed in the inorganic form, whose main sources are sodium selenate (Na_2SeO_4) and sodium selenite (Na_2SeO_3) which provide 42% and 45% of selenium, respectively, or in the organic form as selenomethionine produced by yeasts cultivated in selenium-rich media (Bird et al., 1997).

While sodium selenite is the predominant form of supplementation, the main naturally occurring form in food is L-selenomethionine. Therefore, it is of great importance to differentiate the two sources, inorganic and organic selenium in animal physiology, since bacteria, plants, yeasts or seaweeds are able to synthesize methionine and selenomethionine, but not animals (but, Schrauzer, 2000).

Sodium selenite is normally the most commonly used inorganic source of selenium in dietary supplementation. However, this source can cause oxidative stress, due to the production of peroxide radicals through the reduction reaction with reduced glutathione peroxidase (Surai, 2002).

The formation of free radicals causes the appearance of cardiovascular diseases, cancer, cataracts, muscle degeneration, worsening of carcass quality and a drop in animal production (Surai and Sparks, 2001). Thus, the relationship between antioxidants and pro-oxidants is important in the maintenance of health, embryonic development, productive and reproductive indexes, since there are several factors that govern animal productivity.

Eggs are an important option for solving nutrition-related problems in Latin America, since they are considered a food with complete nutritional value (Souza et al., 2001) and because they have a high biological value, which serves as a standard for measuring the nutritional quality of proteins in other foods (Sakanaka et al., 2000).

Immediately after laying, eggs tend to lose their internal quality if measures are not taken to slow down the rate of the deterioration process (Souza and Souza, 1995). The supplementation of organic selenium in layer diets results in higher egg production and weight, better feed conversion, higher albumen weight and consistency (Rutz et al., 2005), which allows prolonging their storage period.

The concentration of selenium in eggs depends on the concentration and source of selenium in the bird's diet, which ultimately influences embryo development differently between the 10th and 15th day of incubation (Paton et al., 2002).

Organic Se was shown to be more efficient than inorganic Se in improving the performance of the animals, allowing an increase in its concentration in the musculature, which ultimately benefits the product and the final consumer. According to Kim and Mahan (2001), supplementation with organic selenium at increasing levels from 0 to 20 mg/kg promotes a linear deposition of 3.8 to 10.3 mg/kg Se in pork loin.

Selenium is one of the main minerals that act as antioxidants improving male fertility and thus improving reproduction (Alonso et al. 1997). Antioxidants act by protecting the mitochondrial membrane present in spermatozoa which is responsible for motility, fluidity and increased membrane flexibility. However, to perform these functions, mitochondria require high levels of polyunsaturated fatty acids. These high levels leave the cell with increased free radical production and vulnerability to oxidation (Surai, 2002). On the other hand, excess selenium supplementation can cause toxicity. Therefore, depending on the concentration used, selenium can either protect the body against oxidative stress or cause it (Miller et al., 2007).

6.6 Water

Water is a nutritional component of the diet and represents about 75% of the body weight of adult birds and about 65% of the weight of eggs, performing essential functions such as: control of body temperature, aid in digestion and excretion of organic residues (Albino and Barreto, 2003).

Water is classified as a feed with dissolved minerals in its composition, such as calcium, magnesium, sodium chloride, sulfates and bicarbonates. High concentrations of these elements in water are related to low productivity, sudden death, among other diseases that can affect birds (Albino and Barreto, 2003).

Water quality is determined by the absence of bacterial contamination, absence of nitrites and nitrates, as well as absence of organic substances (Albino and Barreto, 2003).

6.7 Examples of quail diet formulations

For illustrative purposes, some examples of diet formulations for quails of double aptitude (fattening and eggs) that provided good productive performance in different production phases are presented in this section (Table 1). It can also be observed some compositions referring to the nuclei (Table 1) and premix (Table 2) that can be acquired commercially. Table 2 shows the calculated composition of a quail diet formulated with corn, soybean bran and soybean oil as sources of energy and protein.

Tables 3 and 4 illustrate the differences that can be observed in the fatty acid composition of feed and egg yolks, respectively, with the inclusion of different types of oils as an energy source in the formulation of diets.

Table 1. Centesimal composition of diets formulated for quail (*Coturnix coturnix coturnix coturnix*) (Roll et al., 2016).

Ingredients	Initial	Growth	Putting
Corn kernels (ground)	41,31	60,27	59,91
Soy flour	52,11	36,28	33,12
Soybean oil	3,31	0,34	2,49
Limestone	1,07	1,08	5,33
Bicalcium phosphate	0,94	1,01	1,32
Core	0,50	0,50	0,50
Salt	0,25	0,25	0,33
DL-methionine	0,33	0,19	-
DL-threonine	0,16	0,07	-
L-Lysine	0,02	-	-

*Levels of guarantees per kg of product: folic acid: 16.7 mg. pantothenic acid: 204,6 mg; zinc bacitracin: 600mg; BHT: 700mg; biotin: 1,4 mg; calcium: 197,5mg; cobalt: 5,1 mg; coper: 244 mg; choline: 42g; iron: 1695mg; (maximum): 400 mg; phosphorus: 50g; iodine: 29mg; magnesium: 1485mg; methionine: 11g; niacin: 840mg; selenium: 3.2mg; sodium: 36g; vitamin A: 2070000IUI; vitamin B1: 40mg; vitamin B12: 430mcg; vitamin B2: 120mg; vitamin B6: 54mg; vitamin D3: 432000IUI; vitamin E: 540mg; vitamin K3: 51.5 mg; zinc: 1535mg.

Table 2. Centesimal composition of a diet formulated for European quail.

Ingredient	(%)
Corn	48,72
Soybean bran	40,20
Soybean oil	2,40
Limestone	5,20
Common salt	0,40
Premix *	3,00

Kaolin	0,08	

Calculated nutritional levels

Metabolizable energy (kcal / kg)	2780	
Crude protein (%)	22,0	
Calcium (%)	2,70	
Available phosphorus (%)	0,46	
Total amino acids (%)	0,74	
Total methionine (%)	0,38	
Total lysine (%)	1,28	
Total cystine (%)	0,36	
Total choline (mg / kg)	2,04	
Linoleic acid (%)	2,60	
Crude fat (%)	4,78	
Crude fiber (%)	3,80	
Sodium (%)	0,20	

Adapted from Roll et al. (2016).

Comparison between the percent fatty acid composition of diets formulated with soybean oil and canola oil for quail (*Coturnix coturnix coturnix coturnix*), following the example diet composition in Table 2.

Type of oil

Dietary fatty acids	Soybeans	Canola
Arachidonic Acid	0,47	0,69
Bhenenic Acid	0,49	0,51
Cis-eicosanoic acid	0,36	0,77
Cis-eicosadienoic acid	0,11	0,09
Stearic Acid	3,59	3,57
Linoleic Acid	49,38	32,38
Linolenic Acid	3,43	2,79
Oleic Acid	26,16	40,18
Palmitic Acid	11,73	10,99
Palmitoleic Acid	0,09	0,18

Adapted from Roll et al. (2016).

Comparison between the percent fatty acid composition of egg yolks from quail fed diets formulated with soybean oil and canola oil for quail (*Coturnix coturnix coturnix coturnix*), following the example diet composition in Table 2.

	Type of dietary oil	
Fatty acids in the yolk	**Soybeans**	**Canola**
Arachidonic Acid	2,05	1,93
Stearic Acid	10,88	10,00
Linoleic Acid	12,5	9,8
Linolenic Acid	0,27	0,30
Oleic Acid	42,24[b]	45,47[a]
Palmitic Acid	24,14	23,55
Palmitoleic Acid	2,36	2,65

Adapted from Roll et al. (2016).

7. Anatomy of the quail

The digestive system of the quail has as components the mouth, esophagus, crop, proventriculus, gizzard, small intestine, caecum, large intestine and cloaca.

1.   *Beak*: responsible for food apprehension, defense and attack.

2.   *Esophagus*: it is a portion of the digestive tract in the form of a tube about 4-5 cm long. Its function is the communication between the pharynx and the proventriculus.

3.   *Buche*: this is a dilatation that divides the esophagus into two portions, one cranial and the other ventral or caudal. The crop, in addition to serving as a food storage bag, also promotes a previous softening of the food that will be sent to the proventriculus.

4.   *Proventriculus*: known as the glandular stomach of birds, it represents a compartment of the digestive tract that is responsible for initiating the gastric digestion of ingested food.

5.   *Gizzard*: this is a muscular organ, also known as the muscular stomach of birds, whose main function is to grind food, reducing it into smaller particles and facilitating digestion. This organ is more developed in herbivorous species due to the higher consumption of fibers, since in this case the food remains for a longer period of time in the compartment.

6.   *Small intestine*: comprising the duodenum, jejunum and ileum, it is the longest portion of the digestive tract and is responsible for most of the digestion and absorption of nutrients from food. A sphincter called pylorus communicates between the small intestine and the gizzard and regulates the passage of digested food between the two portions. The duodenum surrounds the pancreas, which is one of the bird's adnexal glands.

7.   *Large intestine:* in this compartment occurs the final digestion of food being a very short portion constituted by the cecum, colon and rectum. The

cecums are two blind-bottomed pouches located at the transition between the ileum and the colon.

8. *Cloaca:* It is a common compartment, where the digestive, reproductive and urinary systems flow. The cloaca is divided into three parts: the urodeum, where the ureters end, the coprodeum, where the vagina (in females) and vas deferens (in males) end, and the protodeum, where the colon and the rectum end.

The adnexal glands, liver and pancreas, although not part of the digestive system of birds, contribute with secretions to facilitate the digestion of food.

The liver has three main functions:

a)   Inhibit the toxic action of some substances;

b)   Transform carbohydrates and sugars into glycogen, which is transferred to the blood circulation according to the needs of the bird's organism;

c)   To secrete bile to emulsify fats, making them available to lipase enzymes that will make food more absorbable.

The pancreas has two functions, the first is the production of insulin, which in turn is responsible for the regulation of carbohydrate metabolism, and the second is the production of pancreatic juice, which plays an important role in the digestion of food.

7.1 Female reproductive system

When in production activity, the reproductive apparatus of the female can represent about 10% of its live weight, and is made up of two portions, the ovary and the oviduct.

The ovary is attached to the dorsal wall of the bird's body and contains, when the bird is in productive activity, a variety of follicles of different sizes and stages of development that allow for constancy in egg production. The follicles are composed of yolk enveloped by the pre yolk membrane.

The oviduct is made up of several segments, each of which plays an important role in egg formation, and leads to the cloaca. The parts of the oviduct are:

Infundibulum: responsible for receiving the egg that descends from the ovary and storing the sperm.

Magno: responsible for about 90% of albumen formation.

Isthmus: responsible for about the remaining 10% of the albumen and also for the formation of the shell membranes.

Calciferous chamber or shell gland: in this compartment the egg remains most of the time necessary for egg formation, between 16 and 20 hours, for the formation and calcification of the egg shell to occur.

Vagina: after being formed, the egg remains for a short period of time so that it can receive the cuticle of the shell that will promote protection against the entry of microorganisms into the egg.

Cloaca: segment responsible for the pigmentation of the shell and subsequent oviposition, i.e., promoting the passage of the egg to the outside.

Most quail each day normally start laying their eggs from 15:00 hours and close around 23:00 hours, with peak laying between 16:00 and 20:00 hours.

7.2 Male reproductive system

The male reproductive apparatus is responsible for the production of spermatozoa and is composed of two testicles, vas deferens, genital papilla and paragenital glands.

Testicles : located in the abdominal cavity, just below the kidneys.

Vas deferens: located at the end of each vas deferens, they are responsible for the conduction of spermatozoa to the paracloacal glands and storage of semen until the moment of copulation.

Genital papilla: located in the ventral part of the cloaca, it is the copulatory organ of the male.

Paracloacal glands: located above the cloaca, these two glands form an organ that secretes substances that secrete a foamy liquid rich in electrolytes that promote the dilution of semen before copulation and thus increase the fertilizing capacity of the male.

8.   Production factors related to quail welfare.

8.1 Caloric stress

The increase in ambient temperature has a direct effect on poultry and can result in performance losses and reduced egg production, resulting in a series of damages in the production sector.

As a first impact of the increase in temperature, birds reduce feed consumption. Other physiological alterations can also arise, such as, for example, the reduction of the ovarian blood flow and the increase of the respiratory frequency, which aim at increasing the peripheral blood flow to increase heat loss. Along with these two examples, blood circulation and digestive activity of the gastrointestinal tract are also reduced.

Heat stress can be measured through corticosterone levels in birds. Increased corticosterone levels are also a nonspecific response, which is responsible for the production of glucose from substances other than carbohydrates, mainly proteins. These changes are beneficial to the birds in the short term. However, if maintained over a long period of time it will be detrimental with pathological consequences. In general, chronic stressors act by constantly sequestering energy from production (Zulkifli and Siegel, 1995).

Caloric stress can be defined as a set of reactions provoked in the organism due to physical, psychic, infectious and other aggressions that provoke an imbalance in the internal homeostasis of the organism. This disturbance of homeostasis is caused by a series of functional systems of internal control involving physical mechanisms and behavioral reactions.

Thus, as physical and behavioral mechanisms we can cite a series of alterations observed in birds:

a)   Reduction of feed consumption;

b)   Reduction of blood circulation in the gastrointestinal tract;

c)   Reduction of ovarian blood flow;

d)   Reduced digestive activity of the gastrointestinal tract;

e)   Increased peripheral blood flow to increase heat loss;

f)   Increased respiratory rate;

g)   Reduced activity of the enzyme carbonic anhydrase which, in turn, is directly involved in the formation of calcium carbonate ions (important for eggshell formation);

h)   Reduction of the formation of ionized calcium, which is the free form of calcium and is utilized by birds.

The heterophil/lymphocyte ratio is another evaluation tool that can be used as an indicator of stress in birds. Stress when in chronic situation, that is, lasting for long periods, promotes an increase in the number of heterophils with a consequent reduction in the number of lymphocytes in the blood of birds.

In addition, in stress situations there is a greater degradation of proteins at muscular level together with a reduction in protein synthesis, allied to an increase in the mobilization of the birds' body fat reserves.

All these factors and alterations together result in an increase in immunosuppression, i.e., birds present with reduced immunity and can be attacked by a series of opportunistic diseases.

In this sense, the negative impacts of heat stress will not only result in reduced consumption, but will also have a physiological impact by reducing the production and development of the birds.

8.1.1 Negative impacts caused by caloric stress

Reduced egg production and egg weight - the result of reduced nutrient intake and utilization by the birds. In this sense, the contribution of nutrients to the ovary and the weight gain of the quails will also be reduced, as well as a lower production of FSH and LH hormones (which are hormones linked to the reproductive system).

Reduced eggshell quality - results from reduced consumption and utilization of nutrients and lower plasma concentration of ionized calcium, which is responsible for eggshell formation, since it is found in free form, which is better utilized by the birds.

Reduction in weight gain - this reduction is mainly observed in the final phase of poultry rearing, where the birds are more sensitive to temperature changes. The reduction in weight gain is the result of the catabolism of muscle proteins and the reduction of protein synthesis, in addition, of course, to the decrease in nutrient consumption and utilization by the birds. In addition, there is the higher energy expenditure of the birds due to hyperventilation and the reduction of basal metabolism resulting from the decrease in thyroid hormone concentrations.

Increased mortality cases - the result of hyperthermia with consequent physical exhaustion of the bird due to increased respiratory movements, in addition to possible complications of opportunistic diseases arising due to immunosuppression.

8.1.2 Prevention of the negative effects of caloric stress

In order to reduce and even avoid the negative effects of heat stress on poultry production and performance, producers and technicians can try to invest in:

a)   Improvements in the environment of the facilities;

b)   Control the birds' drinking water;

c)   Adequate nutrition in summer periods;

d)   Correct electrolyte balances;

e)   Adjust the housing density according to the period of the year, sex and weight of the birds.

8.2 Beak trimming

Quail, as well as turkeys and chickens, are predisposed to aggressive reactions, such as pecking and cannibalism, and such behaviors become more evident when birds are housed in high stocking densities (Leandro et al., 2005).

To minimize the effects of aggressive behavior in quail, mainly in laying quail (japonica), producers are adopting beak trimming as a management practice, similar to commercial layer rearing (Leandro et al., 2005).

Beak trimming when done correctly does not cause suffering to the animals and does not harm feed and water consumption during the entire production cycle.

On the other hand, beak trimming by untrained and inexperienced people results in damage to the producer, as well as affecting the welfare of the birds.

To perform beak trimming correctly, it is important to consider the use of skilled labor, that is, trained and experienced people who are not in a hurry to perform the procedure, thus guaranteeing a job well done. It is also important to observe other aspects, such as:

Timing of the procedure - the best time to perform beak trimming is at dawn or dusk, as temperatures are usually cooler during these periods, thus avoiding heat stress in the birds. It is also important to keep good quality, fresh water available to the animals at all times.

Sick birds - should never be depopulated and should first be treated until recovery.

Containment of the bird - the bird must be contained calmly and correctly, the technician's index finger must be positioned on the bird's throat promoting the retraction of the tongue thus avoiding the cutting of the same.

The temperature of the trimming blade - (Figure 26) should be around 650 °C before the start of the trimming procedure. Excessively hot blades can cause the formation of neuromas in the beak, which in turn cause discomfort due to increased sensitivity, thus reducing the birds' productive performance.

Figure 25. Despique with use of hot blade

In order to reduce possible problems with the depopulation of the birds, some practices can be carried out to reduce stress, weight loss and feed consumption reduction. Among these practices can be mentioned: a) the supply of vitamin K in the diet for a period of three days before and after the weaning, thus minimizing the possible occurrence of hemorrhages; b) the supply of feed with the feeders well filled so that the birds do not touch the bottom of the feeder with their beaks; c) stimulate the consumption of water and feed through the movement of feed in the feeders; d) avoid in the week after weaning other management practices that can cause some type of stress in the birds.

As positive points with respect to productive performance in flocks of birds that had their beaks trimmed, the increase in egg production rate and the reduction of mortality stand out. These results possibly occur due to the reduction of chopped eggs, reduction of cannibalism and reduction of feed wastage, due to less feed selection by the bird.

Although there are some alternative methods to the hot blade for beak trimming, such as infrared, laser, cold blade or natural attrition with the use of files or plates in the feeders, the conventional method of beak trimming remains the most practical and cost-effective option to improve flock performance and reduce the incidence of cannibalism problems among birds.

9. Bibliography

Aggrey, S.E.; Ankra-Badu, G.A.; Marks, H.L. Effect of longterm divergent selection on growth characteristics in Japanese quail. Poultry Science, v.82, p.538-542, 2003.

Akyurek, H.; Okur, A.A. Effect of storage time, temperature and hen age on egg quality in free-range layer hens. Journal Animal Veterinary Advantage, v.8, p.1953-1958, 2009.

Albino, L.F.T; Barreto, S.L.T. Codornices: criaçâo de codornices para produçâo de ovos e carne. Viçosa: Aprenda Fâcil, 2003, 289p.

Albuquerque, R.; Faria, D.E.; Junqueira, O.M.; et al. Effects of energy level in finisher diets and slaughter age of the performance and carcass yield in broiler chickens. Revista Brasileira de Ciência Avicola, v.5, n.2, p.99-104, 2003.

Alonso, M.L.; Miranda, M.; Hernandez. J.; Castillo, C.; Benedito, J.L. Glutathione peroxidase (GSH-Px) in pathologies associated with selenium deficiencies in ruminants. Archivo de Medicina Veterinària, v.29, n.2,1997.

Angelis, R.C. New concepts in nutrition. Reflexoes a respeito do elo dieta e saúde. Arquivo Gastroenterol-ARQGA/998, v.38, n.4, 2001.

Baiâo, N.C.; Lùcio, C.G. Nutriçâo de matrizes pesadas. In: Macari. M.; Mendes, A.A. Manejo de matrizes pesadas. Campinas: Facta. 2005, cap.10, p.198216, 2005.

Baiâo, N.C; Lara, L.J.C. Oil and fat in broiler nutrition. Revista Brasileira de Ciência Avicola, v.7, n.3, p.129-141, 2005.

Banerjee, S. Carcass studies of Japanese Quails (*Coturnix coturnix japônica*) reared in hot and humid climate of Easter India. World Applied Sciences Journal, v.8, n.2, p.174-176, 2010.

Barreto, S.C.S.; Zapata, J.F.F.; Freitas, E.R.; Fuentes, M.F.F.; Nascimento,

R.F.; Araujo, R.S.R.M; Amorim, A.G.N. Acidos graxos da gema e composiçâo do ovo de poedeiras alimentadas com raçôes com farelo de coco. Pesquisa Agropecuaria Brasileira, v.41, n.12, p.1767-1773, 2006.

Baumgartner, J. Japanese quail production. breeding and genetics. World's Poultry Science Journal. Oxford, v.50, n.3, p. 227-235, 1994.

Belo, M.T.S.; Cotta, J.T.B.; Oliveira, A.I.G. Niveis de energia metabolizâvel em raçôes de codornices japónicas (*Coturnix coturnix japonica*) na fase inicial de puesta. Ciência Agrotécnica, v.24, n.3, p.782-793, 2000.

Bernardino, M.P. Influência dos lipideos da dieta sobre o desenvolvimento ósseo de frangos de corte. Revista Eletrônica Nutritime, v.6, n.3, p.960-966, 2009.

Bertechini, A.G. Nutriçâo de monogàstricos. Lavras: UFLA.2006. 301p.

Brake, J.; Walsh, T.J.; Benton Jr., C.E.; Petitte, J.N.; Meijerhof, R. and Pen Alfa, G. Egg handling and storage. Poultry Science, v.76, p.144-151, 1997.

Briz, R.C. Eggs with higher levels of Ômega 3 fatty acids. In: Simpósio Técnico de Produçâo de Ovos. 1997. Sâo Paulo. Anais... Sâo Paulo: APA. 1997. p.153-193, 1997.

Butolo, J. E. Qualidade de ingredientes na alimentaçâo animal. 1ª Ed. Campinas: Agros Comunicaçâo, 154p., 2002.

Carvalho, P.R.; Pita, M.C.G.; Piber Neto, E.; Mendonça Junior, C.X. Influência da adiçâo de fontes marinhas ricas em PUFAs na dieta sobre a composiçâo lipidica e percentuais de incorporaçâo de PUFAs n-3 na gema do ovo. Arquivos do Instituto Biológico. v.76, n.1, p.27-39, 2009.

Castelo Llobet, J.A, Pontes, M., Franco Gonzalez, F. Egg production. Barcelona: Real Escuela de Avicultura, 1989. 367p.

D'Mello, J.P.F. Amino acids in animal nutrition. 2nd Ed. Wallingford: CABI Publishing, 2003. 546p.

Dolz, S. Utilization of fats and by-products in monogastric animals. Anais...Fundación Espanola para el Desarrollo de la Nutrición Animal - FEDNA. Madrid: Ediciones Peninsular, p.25-38,1996.

Faitarone, A.B.G. Dietary supplementation of lipid sources in dairy cattle and its effects on performance, egg quality, fatty acid profile and cholesterol in the gem. 2010. Tese. 108f. Programa de Pôs-graduaçâo em Zootecnia. Universidade Estadual Paulista. Botucatu, Sâo Paulo, 2010.

Fasenko, G.M. Candling and hatch residue breakouts. In: Robinson. F.E.; Fasenko. G.M.; Renema. R.A. (Eds). Optimizing chick production in broiler breeders. Canada: Spotted Cow, p.101-104, 2003.

Flemming, J.S. Utilizaçâo de leveduras, probióticos mananoligossacarideos (MOS) na alimentaçao de frango de corte. 2005. 109f. Tese (Doutor em Tecnologia do Alimento). Setor de Tecnologia, Universidade Federal do Paranà - Curitiba, 2005.

Fletcher, D.L.; Britton, W.M.; Pesti, G.M.; Rahn, A.P. The relationship of layer flock age and egg weight on egg component yields and solids content. Poultry Science, v.62, p.1800-1805, 1983.

Fujikura, W.S. A posiçâo de Sâo Paulo no mercado nacional de ovos de codorniz e o perfil do consumidor paulistano. In: II Simpósio Internacional e I Congresso Brasileiro de Coturnicultura, 2004, Lavras. MG. Anais...Lavras, 2004.

Garcia, A.R.; Batal, A.B.; Bakert, D.H. Variations in the digestible lysine requirement of broiler chickens due to sex. performance parameters. Rearing environment and processing yield characteristics. Poultry Science, v.85, p.498-504, 2006.

Garcia, E.A.; Mendes, A.A.; Pizzolante, C.C.; Veiga, N. Morphological alterations and performance of quails treated with different feeding programs during the forced molting period. Brazilian Journal of Poultry Science.

Campinas, v.3, p.275-282, 2001.

Gomes, M.R.; Tirapegui, J. Relationship between some nutritional supplements and physical performance. Archivo Latinoamericano Nutriçâo, v.50, n.4, p.317329, 2000.

Hirayama, K.B.; Patricia, G.L.; Speridiâo, P.G.L.; Fagundes Neto U. Acidos graxos polinsaturados de cadeia longa. The Eletronic Journal of Pediatric Gastroenterology. Nutrition and Liver Diases, v.10, n.3, 2006.

Adolfo Lutz Institute (IAL). Normas analiticas do Instituto Adolfo Lutz: Métodos fisico-quimicos para anàlise de alimentos. 4ª ed., 1ª Ediçâo Digital, Sâo Paulo, 1020p., 2008.

IBGE - Instituto Brasileiro de Geografia e Estatistica. Diretoria de Pesquisas, Coordenaçâo de Agropecuària, Pesquisa da Pecuària Municipal 2013.

IBGE - Brazilian Institute of Geography and Statistics. Produçâo da Pecuària Municipal. Rio de Janeiro, v.42, p.1-39, 2014.

Jones, J.E.; Hughes, B.L.; Hale, K.K. Coturnix D1 carcass yield. Poultry Science, v.58, p.1647-1648, 1979.

Leandro, N.S.M.; Stringhini, J.H.; Café, M.B. et al. Efeito da granulometria do milho e do farelo de soja sobre o desempenho de codornices japonicas. Revista Brasileira de Zootecnia, v.30, n.4, p.1266-1271, 2001.

Leandro, N.S.M.; Vieira, N.S.; Matos, M.S.; Café, M.B.; Stringhini, J.H. e Santos, D.A. Desempenho produtivo de codornices japónicas (*Coturnix coturnix japonica*) submetidas a diferentes densidades e tipos de debicagem. Acta Scientiarum. Animal Sciences, v.27, n,1, p.129-135, 2005.

Lourens, A.; Molenaar, R.; Van Den Brand, H.; Heetkamp, M.J.; Meijerhof, R. and Kemp, B. Effect of egg size on heat production and the transition of energy from egg to hatchling. Poultry Science, v.85, p.770-776, 2006.

Longo, F.A.; Menten, J.F.M.; Pedroso, A.A.; Figueiredo, A.N.; Racanicci,

A.M.C.; Galotto, J.B.; Sorbara, J.O.B. Diferentes fontes de proteina na dieta pré- inicial de frangos de corte. Revista Brasileira de Zootecnia, v.34, n.1, p.112-122, 2005.

Macari, M.; Furlan, R.L.; Gonzalez, E. Fisiologia aviària aplicada a frangos de corte. Jaboticabal: Universidade Estadual Paulista, 2002.

Mandal, A.B.; Elangovan, A.V.; Tyagi, P.K.; Tyagi, P.K.; Tyagi, A.K.J.; Kaur, S. Effect of enzyme supplementation on the metabolizable energy content of solvent-extracted rapeseed and sunflower seed meals for chicken, guinea fowl and quail. British Poultry Science, v.46, p.75-79, 2005.

Mandal, A.B.; Kaur, S.; Johri, A.K.; Elangovan, A.V.; Deo, C.; Shrivastava, H.P. Response of growing Japanese quails to dietary concentration of L-threonine. Journal of the Science and Food and Agriculture, v.86, p.793798, 2006.

Marks, H.L. Feed efficiency changes accompanying selection for body weight in chickens and quail. World's Poultry Science, v.47, p.197-212, 1991.

Menge, H. Linoleic acid requirement of the hen for reproduction. Journal of Nutrition; v.95, p.578-72, 1968.

Minvielle, F. The future of Japanese quail for research and production. World Poultry Science Journal, v.1, p.500-507, 2004.

Minvielle, F.; Oguz, Y. Effect of genetics and breeding on egg quality of Japanese quail. World's Poultry Science Journal, v.58, p.291-295. 2002.

Moraes, M.A.C. Métodos para avaliaçâo sensorial dos alimentos. 5ª Ed. Campinas: Unicamp, 85p., 1985.

Móri, C.; Garcia, E.A.; Pavan, A.C.; Piccinin, A.; Scherer, M.R.; Pizzolante, C.C. Desempenho e qualidade dos ovos de codornices de quatro grupos genéticos. Revista Brasileira de Zootecnia, v.34, n.3, p.864-869, 2005.

Morita, M.M. Custo x beneficio do uso de óleos e gorduras em dietas avicolas. In: Apinco Conference on Poultry Science and Technology. 1992. Santos. Anais...Santos: Apinco, p.29-35, 1992.

Murakami, A.E.; Furlan, A.C. Research on nutrition and feeding of laying quails in Brazil. In: Simpósio Internacional de Coturnicultura. 2002. Lavras. MG. Anais... Lavras: Universidade Federal de Lavras. p.113-120. 2002.

Murakami, K.T.T. Óleo de linhaça como principal fonte lipidica na dieta de frangos de corte. Araçatuba. 2009.   64f. Dissertaçâo (Mestrado). Paulista State University. School of Dentistry and Veterinary Medicine Course, 2009.

National Research Council - NRC. Nutrient requirements of poultry. Washington: National Academy of Sciénces, p.44-45, 1994.

Nowaczewski, S.; Kontecka, H.; Rosinski, A.; Koberling, S.; Koronowski. P. Egg Quality of Japanese Quail Depends on Layer Age and Storage Time. Folia biologica (Kraków), v.58, p.201-207, 2010.

Noy, Y.; Sklan, D. Digestion and absorption in the young chick. Poultry Science, v.74, p.366-373, 1995.

Noronha, J.F. Apontamentos de Anâlise Sensorial, Escola Superior Agrària de Coimbra (ESAC), 2003. 75p.

Nunes, I.J. Nutriçâo animal bàsica. 2Ed. Belo Horizonte: FEP-MVZ, 1998. 383p.

Paton, N.D.; Cantor, A.H; Pescatore, A.J.; Ford, M.J.; Smith, C.A. Absorption of selenium by developing chick embryos during incubation. In: Lyons, T.P.; Jacques, K.A. Biotechnology in the Feed Industry. 18th Alltech's Annual symposium. Proceedings... Nottingham, UK: Nottingham University Press, p.107-121, 2002.

Pinto, R.; Ferreira, A.S.; Albino, L.F.T.; Gomes, P.C., Vargas Jr., J.G. iveis de proteina e energia para codornices japónicas em puesta. Revista Brasileira de Zootecnia, v.32, n.5. p.1761-1770, 2002.

Pita, M.C.G.; Piber Neto, E.; Carvalho, P.R.; Mendonça Jr., C.X. Efeito da suplementaçâo de linhaça, óleo de canola e vitamina E na dieta sobre as concentraçôes de ácidos graxos poliinsaturados em ovos de galinha. Arquivo Brasileiro Medicina Veterinària Zootecnia, v.58, n.5, p.925-931, 2006.

Reis, L.F.S.D. Codornizes, criaçâo e exploraçâo. Lisbon, Agros, 10, 1980. 222p.

Redder, E. Compare quail eggs with chicken eggs. Revista Saùde é vital. Sâo Paulo, p.27, April, 2005.

Reis, J.S. Caracteristicas da carcaça de uma linhagem de codornices de corte. 2011. 88f, Dissertaçâo (Mestrado). Programa de Pôs-Graduaçâo em Zootechnics. Universidade Federal de Pelotas, 2011.

Rezende, M.J.M.; Flauzina, L.P.; Mcmanus, C.; Oliveira, L.Q.M. Desempenho produtivo e biometria das visceras de codornices francesas alimentadas com diferentes niveis de energia metabolizàvel e proteina bruta. Acta Scientiarum. Animal Sciences, v. 26, n. 3, p. 353-358, 2004.

Ribeiro, T.C.; Moreira, P.C.; Oliveira, J.P.; Lemos, E.N.; Silva, S.M. Influência do peso, à incubaçâo, na eclodibilidade de ovos de avestruz. Estudos-Goiânia, v.35, n.3, p.501-516, 2008.

Rocha, J.S.R.; Lara, L.J.C.; Baiâo, N.C.; Cançado, S.V; Baiâo, L.E.C; Silva, T.R. Efeito da classificaçâo dos ovos sobre o rendimento de incubaçâo e os pesos do pinto e do saco vitelino. Arquivo Brasileiro Medicina Veterinària Zootecnia, v.60, n.4, p.979-986, 2008.

Rodrigues, E.A.; Cancherini, L.C.; Junqueira, O.M.; De Laurentiz, A.C.; Silva Filardi, R.; Duarte, K.F.; Casartelli, E.M. Performance, shell quality and lipid profile of egg gems from commercial broilers fed increasing levels of soybean oil in the second laying cycle. Acta Scientiarum Animal Sciences, v.27, n.2, p.207-212, 2005.

Roll, A.A.P. Canola oil and organic selenium for dual purpose quail. Pelotas, Faculdade de Agronomia Eliseu Maciel da Universidade Federal de Pelotas, 2012. 84f. Dissertaçâo (Mestrado em Zootecnia) - Faculdade de Agronomia Eliseu Maciel da Universidade Federal de Pelotas, 2012.

Roll, A.A.P.; Hobuss, C.B.; Del Pino, F.A.B.; Roll, V.F.B.; Dionello, N.J.L.; Xavier, E.G.; Rutz, F. Canola oil and organic selenium in quail diets: fatty acid profile, cholesterol content and external egg quality. Semina: Ciências Agrârias, v.37, n.1, p.405-414, 2016.

Roll, V.F.B.; Cepero, R.C.; Levrino, G.A.M. Floor versus cage rearing: effects on production. egg quality and physical condition of laying hens housed in furnished cages. Ciência Rural, v.39, n.5, 1527-1532, 2009.

Rostagno. H.S. Tabelas brasileiras para aves e suinos: Composiçâo de Alimentose Exigências Nutricionais. 2ª ediçâo. Viçosa: UFV.

Department of Animal Husbandry. 186p. 2005.

Sakamoto, M.I.; Murakami, A.E.; Souza, L.M.G.; Franco, J.R.G.; Bruno, L.D.G.; Furlan, A.C. Energy value of some alternative feeds for japonica quails. Revista Brasileira Zootecnia, v.35, n.3, p.818-821, 2006.

Santos. J.E.C.; Gomes. F.S.; Borges. G.L.F.N. et al. Efeito da linhagem e da idade das matrizes na perda de peso dos ovos e no peso embrionàrio durante a incubaçâo artificial. Bioscience Journal, v.25, p.163-169, 2009.

Scragg, R.H.; Logan, N.B.; Guedes, N. Response of egg weight to the inclusion of fat in layer diets. British Poultry Science, v.28, p.15-21, 1987.

Seibel, N.F.; Schoffen, D.B.; Queiroz, M.I.; Almeida de Souza-Soares, L. Sensory characterization of quail eggs fed modified diets. Ciência e Tecnologia de Alimentos, v.30, n.4, p.884-889, 2010.

Silva, E.L.; Silva, J.H.V.; Jordâo Filho, J.; Ribeiro, M.L.G.; Costa, F.G.P.; Rodrigues, P.B. Reduçâo dos niveis de proteina e suplementaçâo aminoacidica em rações para codornices européias (*Coturnix coturnix coturnix coturnix*). Revista Brasileira de Zootecnia, v.35, n.3, p.822-829, 2006.

Silva. J.H.V.; Jordâo Filho. J.; Costa, F.G.P; Lacerda, P.B.; Vargas, D.G.V. Nutritional requirements of quails. In: XXI Congresso Brasileiro de Zootecnia. 2011. Maceió. AL. Anais... Maceió: Universidade Federal de Alagoas, CD-ROM, 2011.

Singh, R.P.; Panda, B. Effect of seasons on physical quality and component yields of egg from different lines of quail. Indian Journal of Animal Science, v.57, n.1, p.50-55, 1987.

Soares, H.F.; Ito, M.K. O àcido graxo monoinsaturado do abacate no controle das dislipidemias. Revista Ciência Médica, v.9, n.2, p.47-51, 2000.

Souza, P.A.; Souza, H.B.A.; Oba, A.; Gardini, C.H.C. Influence of ascorbic acid on egg quality. Ciência e Tecnologia de Alimentos, v.21, n.3, p. 273-275, 2001.

Stone, H.; Sidel, J.L. Quantitative Descriptive Analysis: Developments, Applications, and the Future. Food Techology, v.52, n.8, p.48-52, 1998.

Surai, P.F.; Sparks, N.H.C. Designer eggs: from improvement of egg composition to functional food. Trends in food science and technology, v.7: p.12-16, 2001.

Surai, P.F. Natural Antioxidants in Avian Nutrition and Reproduction. 1ª Ed. Nottingham University Press, Nottingham, p.27-95, 2002.

Thomazini, M.; Franco, M.R.B. Metodologia para anâlise dos Constituintes volâteis do sabor. Boletim SBCTA, v.34, n.1, p.52-59, 2000.

Valenzuela, A.B.; Sanhueza, J.C. Oils of marine origin; their importance in nutrition and food science. Revista Chilena Nutricion, v.36, n.3, 2009.

Valenzuela, A.B.; Nieto, S.K. Omega-6 and omega-3 fatty acids in perinatal nutrition: their importance in the development of the nervous and visual system. Revista Chilena Pediatria, v.74, p.149-57, 2003.

Vieira. S.L.; Moran Jr., E.T. Effects of egg of origin and chick post-hatch nutrition on broiler live performance and meat yields. World's Poultry Science Journal, v.55, p.125-142, 1999.

Villela. J.L. Criaçâo de codornices. Coleçâo Agroindùstria, v.14, 91p. Cuiabà: SEBRAE/MT. 1998.

Whitehead, C.; Bowman, A.Y.; Griffin, H. The effects of dietary fat and bird age on the weights of eggs and egg components in the laying hen. British Poultry Science, v.32, p.565-574. 1991.

Zulkifli, I.; Siegel, P.B. Is there a positive side to stress? World's Poultry Science Journal 51:63-76, 1995.

10. Authors

Aline Piccini Roll

Post-Doctorate of the Post-Graduate Program in Animal Husbandry at the Federal University of Pelotas - UFPEL.

Mèdica Veterinaria, Ms.C, Dra. en Nutrición Animal por la UFPEL with academic internship abroad at the Universidad de Zaragoza and at the Universitat Autònoma de Barcelona, Spain.

Email: apiroll@yahoo.es

Fernando Rutz

Ph.D. in Animal Nutrition from the University of Kentucky, U.K.Y, United States.

Veterinary Doctor, Ms.C in Animal Nutrition by UFPEL.

Associate Professor of the Animal Husbandry Course at UFPEL.

Email: frutz@alltech.com

Victor Fernando Büttow Roll

Post-Doctorate in Animal Production, Universitat Autònoma de Barcelona, Spain.

D. in Animal Production from the University of Zaragoza - Spain.

Agricultural Engineer, Ms.C. in Animal Nutrition by UFPEL.

Associate Professor of the Animal Husbandry Course at UFPEL.

Email: roll98@ufpel.edu. br

I want morebooks!

Buy your books fast and straightforward online - at one of world's fastest growing online book stores! Environmentally sound due to Print-on-Demand technologies.

Buy your books online at
www.morebooks.shop

Kaufen Sie Ihre Bücher schnell und unkompliziert online – auf einer der am schnellsten wachsenden Buchhandelsplattformen weltweit! Dank Print-On-Demand umwelt- und ressourcenschonend produzi ert.

Bücher schneller online kaufen
www.morebooks.shop

Printed by Books on Demand GmbH, Norderstedt / Germany